2010

LLEWELLYN'S

MOON
SIGN
BOOK

Llewellyn's 2010 Moon Sign Book®

ISBN 978-0-7387-0688-7.

Cover Design & Art: Ellen Dahl
Editor: Nicole Edman
Designer: Sharon Leah
Stock photography model(s) used for illustrative purposes only and may not endorse or represent the book's subject.
Copyright 2009 Llewellyn Worldwide. All rights reserved.
Typography owned by Llewellyn Worldwide.

Any Internet references contained in this work are current at publication time, but the publisher cannot guarantee that a specific location will continue to be maintained.

Astrological data compiled and programmed by Rique Pottenger. Based on the earlier work of Neil F. Michelsen.

You can order Llewellyn annuals and books from *New Worlds*, Llewellyn's catalog. To request a free copy of the catalog, call toll-free 1-877-NEW-WRLD, or visit our Web site at www.llewellyn.com.

Llewellyn is a registered trademark of Llewellyn Worldwide, Ltd.
2134 Wooddale Drive, Woodbury, MN 55125-2989 USA
Moon Sign Book® is registered in U.S. Patent and Trademark Office.
Moon Sign Book is a trademark of Llewellyn Worldwide, Ltd. (Canada).
Printed in the USA

Llewellyn Worldwide
Dept. 978-0-7387-0688-7
2143 Wooddale Drive
Woodbury, MN 55125-3989
www.llewellyn.com

Table of Contents

What's the Moon Got to Do With It?

A Note from Carl Llewellyn Weschcke

From both practical and esoteric points of view, you could easily say the Moon has nearly everything to do with it! The Moon is more than Earth's natural satellite. From Earth, we see Sun and Moon as a natural pair; but from the Sun, Earth and Moon are the natural pair, the yin and yang of human experience. The Moon does not merely reflect the Sun's light onto the Earth's surface but also transforms the energies of the Sun's radiation and energies from other planetary and sidereal bodies. More about that later.

Back in 1959, I flew to Los Angeles and visited the small building in Culver City housing Llewellyn Publications. I then arranged to purchase the business. Llewellyn George, the founder, died in 1954, and the business was operated by a printer. At the time, Llewellyn's most important product was the annual *Moon Sign Book & Gardening Almanac*. The main trade customers were mail order "seed houses," and their customers were mostly farmers and gardeners using the lunar tables to choose the best times for their planting and harvesting activities.

But the value of lunar timing is not limited to better garden and farm-related activities. All sorts of biological and human activities benefit from making the proper timing decisions as prescribed by the Moon's signs and phases. You could say: "Better Living by the Light of the Moon."

The Moon is a kind of "transformer" for all sorts of astronomical energy sources. It receives light and energy from the Sun, the planets of our solar system, and other cosmic sources, and relays those energies to Earth. But the Moon also changes those energies moving through the combined electromagnetic field of Moon and Earth. Of course, the Moon's most familiar effect is the gravitational pull producing the ocean tides, but sophisticated measurements show that same influence affects fluids in very small bodies of water and other fluids—even the fluids in plants and human bodies.

Both the "ancients" and modern scientists recognize the subtle influences of the Moon and of other planetary and ethereal influences through the Moon. There's tradition, verification, and new understanding and application of these factors to meet our contemporary readers' needs.

One of our challenges is to put this knowledge into simple "rules" and describe them in text and convenient tables for easy application to our readers' interests and needs. You're not particularly interested in the complexity of the calculations that go into this work, but you do want to quickly and easily discover the best timing for the activities that are important to you.

There will inevitably be times when events bring about more unusual situations that may require the reader to extrapolate from the known factors and rules to the larger unknown. We try to lay the foundation for that kind of extrapolation.

The Character of the Moment

The entire foundation for electional astrology—and agricultural and horticultural applications are the oldest and most fundamental of all—is that we can make intelligent timing decisions for greater benefit. Instead of being innocent victims of the unknown, we discover the known and work in harmony with natural cycles. While nonagricultural applications are further removed from the more obvious cycles, the principles are the same, and the results equally worthwhile.

Within the last few years, science has demonstrated what the ancients discovered long ago: there is interconnectedness among all things, so that any intentional action is like a "snap-shot" that locks all of those influences together into a new birth, whether of a plant or a baby or a signed contract. The moment in time that something new was born is reflected in a horoscope or birth chart—a measure of that moment in terms of astronomical factors. It is a beginning, and from that moment on, the plant/baby/contract has a new life.

To put it another way, every activity takes place within a "field" that is sensitive to human intention (be it the basic gravitational field, the more pervasive electromagnetic field, or spheres of yet more subtle energies). The more we observe the field and the more we understand it, the more we can work beneficially with it and influence it.

The permeations of this field study give us two kinds of astrology: electional astrology and what I prefer to call event-based astrology. Electional astrology lets us look for the influences operating within a range of times to choose the precise time that most benefits the action we want to take. The event horoscope gives us understanding of the influences actually surrounding the moment of birth whether of a child, a nation, or any other thing.

The Rhythm of the Universe

The underlying aim of the Universe is growth and expansion. We see that impulse toward growth wherever we look. Another name for it is "love." Love is relationship and nourishment. When we act with awareness and intention, we most often are in a loving relationship with plant, animal, life, Nature, and the goodness of a better life. Love brings meaning, purpose, and happiness to our world.

But while growth is the dynamic expansionary force, there is a contrary force of death followed by regeneration. Together, they exist not in equilibrium, but in cyclical growth and a rhythm of expansion. Even our planet, Earth, has grown in size through the generations of birth and decay and rebirth.

We grow, and we become more than we were. We learn to live in, and build upon, harmony with the natural processes, and we "go forth and multiply."

The Secret Book of Nature

Nature is the Book of Life, for Nature is alive and is our source of being. Nature is the Mother who bears and nurtures us, and the Father who teaches and inspires us. Nature houses our body, mind, and spirit. We stand on her Earth and reach to his sky. Nature is an open book but requires effort to uncover her secrets. We start with the Love of Life, the inspiration to grow, the desire to learn, and the need to become more than we are.

—Carl Llewellyn Weschcke

Nature is itself the great teacher. We study, we learn, and then we apply the lessons learned in order to grow and multiply. We grow our crops and harvest both seed to replant and surplus for our own use and to sell. We learn to combine our natural resources with our knowledge of the Moon's signs and phases to improve our soil, to grow healthier plants that resist disease and destructive parasites, and to harvest correctly to preserve our crop over the cold winter months.

We breed our animals and poultry according to principles learned in lunar agriculture, and likewise we increase our herds and flocks, we raise healthier progeny, and the cycle of growth and improvement continues.

Through these lessons, we become co-creators with Nature to fulfill the underlying directive that compels the Universe to grow. Our role is to become active agents for change, growth, and renewal. But, we are also free agents and can work against the natural laws and bring harm and destruction, a loss of life and beauty, and great suffering in our wake.

To live harmoniously and with purpose is the responsibility we're given.

The Handbook of Natural Living

It's also been our goal to make *Llewellyn's Moon Sign Book* the best guide to natural living whether you live in rural or urban or suburban areas. We try to anticipate your needs, whether it's the best date for listing your property for sale or marrying a widow, for pouring concrete or baking a cake.

For 105 years, the *Moon Sign Book* has been a helpful friend to millions of readers. Let's celebrate this long journey. Have you grown the world's largest tomato or the tallest corn using Lunar power? Has the Best Times system worked astonishingly well for you? Share your successes with us and other readers for the joy of true natural living.

Llewellyn Worldwide, Ltd.
Dept. 978-0-7387-0688-7
2143 Wooddale Drive
Woodbury, MN 55125-3989

What's Different About the Moon Sign Book?

Readers have asked why *Llewellyn's Moon Sign Book* says that the Moon is in Taurus when some almanacs indicate that the Moon is in the previous sign of Aries on the same date. It's because there are two different zodiac systems in use today: the tropical and the sidereal. *Llewellyn's Moon Sign Book* is based on the tropical zodiac.

The tropical zodiac takes 0 degrees of Aries to be the Spring Equinox in the Northern Hemisphere. This is the time and date when the Sun is directly overhead at noon along the equator, usually about March 20–21. The rest of the signs are positioned at 30-degree intervals from this point.

The sidereal zodiac, which is based on the location of fixed stars, uses the positions of the fixed stars to determine the starting point of 0 degrees of Aries. In the sidereal system, 0 degrees of Aries always begins at the same point. This does create a problem though, because the positions of the fixed stars, as seen from Earth, have changed since the constellations were named. The term "precession of the equinoxes" is used to describe the change.

Precession of the equinoxes describes an astronomical phenomenon brought about by the Earth's wobble as it rotates and orbits the Sun. The Earth's axis is inclined toward the Sun at an angle of about 23½ degrees, which creates our seasonal weather changes. Although the change is slight, because one complete circle of the Earth's axis takes 25,800 years to complete, we can actually see that the positions of the fixed stars seem to shift. The result is that each year, in the tropical system, the Spring Equinox occurs at a slightly different time.

Does Precession Matter?

There is an accumulative difference of about 23 degrees between the Spring Equinox (0 degrees Aries in the tropical zodiac and 0 degrees Aries in the sidereal zodiac) so that 0 degrees Aries at Spring Equinox in the tropical zodiac actually occurs at about 7 degrees Pisces in the sidereal zodiac system. You can readily see that those who use the other almanacs may be planting seeds (in the garden and in their individual lives) based on the belief that it is occurring in a fruitful sign, such as Taurus, when in fact it would be occurring in Gemini, one of the most barren signs of the zodiac. So, if you wish to plant and plan activities by the Moon, it is helpful to follow *Llewellyn's Moon Sign Book*. Before we go on, there are important things to understand about the Moon, her cycles, and their correlation with everyday living. For more information about gardening by the Moon, see page 65.

Weekly Almanac

Your Guide to Lunar Gardening & Good Timing for Activities

In the end, there is really nothing more important than taking care of the earth and letting it take care of you.
~CHARLES SCOTT

January

January 1–2

In the depth of winter, I finally learned that within me there lay an invincible summer.

~ALBERT CAMUS

Date	Qtr.	Sign	Activity
Dec 31, 2:13 pm–Jan 1, 9:41 pm	3rd	Cancer	Plant biennials, perennials, bulbs, and roots. Prune. Irrigate. Fertilize (organic).
Jan 1, 9:41 pm–Jan 3, 9:52 pm	3rd	Leo	Cultivate. Destroy weeds and pests. Harvest fruits and root crops for food. Trim to retard growth.

Ancient Babylonians observed the New Year celebration as long ago as 4000 BC, but the date changed from year to year. In 153 BC, the Roman Senate declared January 1 the date to celebrate the new year, but various emperors would change the date. In 46 BC, Julius Caesar established the Julian calendar that is still in use today. In order to synchronize the calendar with the Sun, that year was actually 445 days long!

JANUARY

S	M	T	W	T	F	S
					1	2
3	4	5	6	7	8	9
10	11	12	13	14	15	16
17	18	19	20	21	22	23
24	25	26	27	28	29	30
31						

2009 © JinYoung Lee. Image from BigStockPhoto.com

January 3–9

Beauty is how you feel inside, and it reflects in your eye. It is not something physical.

~Sophia Loren

Date	Qtr.	Sign	Activity
Jan 1, 9:41 pm– Jan 3, 9:52 pm	3rd	Leo	Cultivate. Destroy weeds and pests. Harvest fruits and root crops for food. Trim to retard growth.
Jan 3, 9:52 pm– Jan 5, 11:58 pm	3rd	Virgo	Cultivate, especially medicinal plants. Destroy weeds and pests. Trim to retard growth.
Jan 8, 5:00 am– Jan 10, 1:10 pm	4th	Scorpio	Plant biennials, perennials, bulbs, and roots. Prune. Irrigate. Fertilize (organic).

Do deep skin cleansing while the Moon is waning. Easy to use and readily available herbal treatments include peppermint tea and sea salt. A clean cloth soaked in peppermint tea and applied to the face is a good astringent, and a paste made from sea salt and water that is applied to the face and left on for 20 minutes before rinsing off is a refreshing beauty pack.

_____

_____

_____

January 7
5:40 am EST

		JANUARY				
S	M	T	W	T	F	S
					1	2
3	4	5	6	7	8	9
10	11	12	13	14	15	16
17	18	19	20	21	22	23
24	25	26	27	28	29	30
31						

January 10–16

Sad soul, take comfort, nor forget
The sunrise never failed us yet.

~CELIA THAXTER

Date	Qtr.	Sign	Activity
Jan 8, 5:00 am–Jan 10, 1:10 pm	4th	Scorpio	Plant biennials, perennials, bulbs, and roots. Prune. Irrigate. Fertilize (organic).
Jan 10, 1:10 pm–Jan 12, 11:54 pm	4th	Sagittarius	Cultivate. Destroy weeds and pests. Harvest fruits and root crops for food. Trim to retard growth.
Jan 12, 11:54 pm–Jan 15, 2:11 am	4th	Capricorn	Plant potatoes and tubers. Trim to retard growth.
Jan 15, 2:11 am–Jan 15, 12:17 pm	1st	Capricorn	Graft or bud plants. Trim to increase growth.

Long, dark winter days and indoor living reduce our exposure to sunlight and increase our chances of being deficient in vitamin D—the "sunshine vitamin." A vitamin D deficiency is associated with low moods, a common complaint during the winter months. Coincidence? Maybe. But why take a chance when a vitamin supplement or a ten-minute walk in the sunshine may help cure winter doldrums?

January 15
2:11 am EST

JANUARY

S	M	T	W	T	F	S
					1	2
3	4	5	6	7	8	9
10	11	12	13	14	15	16
17	18	19	20	21	22	23
24	25	26	27	28	29	30
31						

2009 © Goce Risteski. Image from BigStockPhoto.com

January 17–23

The time is always right to do what's right.

~MARTIN LUTHER KING, JR.

Date	Qtr.	Sign	Activity
Jan 18, 1:17 am– Jan 20, 1:36 pm	1st	Pisces	Plant grains, leafy annuals. Fertilize (chemical). Graft or bud plants. Irrigate. Trim to increase growth.
Jan 22, 11:39 pm– Jan 23, 5:53 am	1st	Taurus	Plant annuals for hardiness. Trim to increase growth.
Jan 23, 5:53 am– Jan 25, 6:11 am	2nd	Taurus	Plant annuals for hardiness. Trim to increase growth.

In about twelve weeks, April showers will bring May flowers, and now is the time to start your own plants if you do that. Many annuals can be sown directly in the soil when it gets warmer, but some benefit from an earlier start. Pansy, viola, and lobelia have a slower rate of growth, and they'll do well if planted now and transplanted outdoors in the spring when temperatures are still cool. Wait until the week following Valentine's Day to start impatiens and petunias, though.

January 23
5:53 am EST

JANUARY

S	M	T	W	T	F	S
					1	2
3	4	5	6	7	8	9
10	11	12	13	14	15	16
17	18	19	20	21	22	23
24	25	26	27	28	29	30
31						

~~~ January 24–30

The most beautiful thing we can experience is the mysterious.

—ALBERT EINSTEIN

Date	Qtr.	Sign	Activity
Jan 23, 5:53 am– Jan 25, 6:11 am	2nd	Taurus	Plant annuals for hardiness. Trim to increase growth.
Jan 27, 9:01 am– Jan 29, 9:10 am	2nd	Cancer	Plant grains, leafy annuals. Fertilize (chemical). Graft or bud plants. Irrigate. Trim to increase growth.
Jan 30, 1:18 am– Jan 31, 8:23 am	3rd	Leo	Cultivate. Destroy weeds and pests. Harvest fruits and root crops for food. Trim to retard growth.

On January 30, 1958, the first commercial two-way moving sidewalk went into service at Love Field in Dallas, Texas. Today, moving sidewalks are found in airports, museums, at zoos (The aquarium at the Mall of America has one.), and theme parks. A high-speed moving sidewalk (travels at 6 mph) was installed in the Montparnasse—Bienvenue Metro station in Paris in 2002. It has been estimated that if a person used the Bienvenue Metro sidewalk twice a day, they would save 11.5 hours per year.

_____

_____

_____

O

January 30
1:18 am EST

JANUARY

S	M	T	W	T	F	S
					1	2
3	4	5	6	7	8	9
10	11	12	13	14	15	16
17	18	19	20	21	22	23
24	25	26	27	28	29	30
31						

February

January 31–February 6

O, Wind, If Winter comes, can Spring be far behind?
                    ~PERCY BYSSHE SHELLEY

Date	Qtr.	Sign	Activity
Jan 30, 1:18 am– Jan 31, 8:23 am	3rd	Leo	Cultivate. Destroy weeds and pests. Harvest fruits and root crops for food. Trim to retard growth.
Jan 31, 8:23 am– Feb 2, 8:42 am	3rd	Virgo	Cultivate, especially medicinal plants. Destroy weeds and pests. Trim to retard growth.
Feb 4, 11:56 am– Feb 5, 6:48 pm	3rd	Scorpio	Plant biennials, perennials, bulbs, and roots. Prune. Irrigate. Fertilize (organic).
Feb 5, 6:48 pm– Feb 6, 7:04 pm	4th	Scorpio	Plant biennials, perennials, bulbs, and roots. Prune. Irrigate. Fertilize (organic).
Feb 6, 7:04 pm– Feb 9, 5:43 am	4th	Sagittarius	Cultivate. Destroy weeds and pests. Harvest fruits and root crops for food. Trim to retard growth.

Root crops grow best in cool temperatures that range from 50°F to 75°F. Radishes and carrots do best in soil that is about 50°F, while beets are sweeter when grown in 75°F soil.

2009 © Marilyn Barbone. Image from BigStockPhoto.com

February 5
6:48 pm EST

FEBRUARY

S	M	T	W	T	F	S
		1	2	3	4	5	6
7	8	9	10	11	12	13
14	15	16	17	18	19	20
21	22	23	24	25	26	27
28						

~~~ February 7–13

*Enlightenment comes not from teaching but through the eye
awaked inwardly.*

~YASUNARI KAWABATU

Date	Qtr.	Sign	Activity
Feb 6, 7:04 pm– Feb 9, 5:43 am	4th	Sagittarius	Cultivate. Destroy weeds and pests. Harvest fruits and root crops for food. Trim to retard growth.
Feb 9, 5:43 am– Feb 11, 6:24 pm	4th	Capricorn	Plant potatoes and tubers. Trim to retard growth.
Feb 11, 6:24 pm– Feb 13, 9:51 pm	4th	Aquarius	Cultivate. Destroy weeds and pests. Harvest fruits and root crops for food. Trim to retard growth.

Small gardens are appealing to today's gardener, but small doesn't necessarily mean less work. If anything, we need to resist the temptation to add too much variety. The Rule of Three can be helpful to remember when selecting plants: that's three shades of pink, or three plant types, or three variations of something, for example. And keep things in proportion. While it is not necessary to use "small" plants in small spaces, your plantings should be in proportion to the space and surroundings.

February 13
9:51 pm EST

FEBRUARY

S	M	T	W	T	F	S
			1	2	3	4	5	6
7	8	9	10	11	12	13
14	15	16	17	18	19	20
21	22	23	24	25	26	27
28						

February 14–20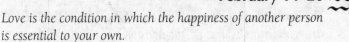

Love is the condition in which the happiness of another person is essential to your own.

~Robert A. Heinlein

Date	Qtr.	Sign	Activity
Feb 14, 7:23 am– Feb 16, 7:30 pm	1st	Pisces	Plant grains, leafy annuals. Fertilize (chemical). Graft or bud plants. Irrigate. Trim to increase growth.
Feb 19, 5:55 am– Feb 21, 1:47 pm	1st	Taurus	Plant annuals for hardiness. Trim to increase growth.

Like a bouquet of roses with a touch of baby's breath, a garden that relies on the repetition of a few varieties and colors often has a bigger impact than does a garden that overwhelms the eye with too many shapes, colors, and textures. Single colors or repeated groupings can create a unified, simple design that is both pleasing and restful to look at. Mass plantings are dramatic—especially when they complement large rocks, stone paths, or ornamental trees in the garden.

_____

_____

_____

_____

FEBRUARY						
S	M	T	W	T	F	S
	1	2	3	4	5	6
7	8	9	10	11	12	13
14	15	16	17	18	19	20
21	22	23	24	25	26	27
28						

2009 © V. J. Matthew. Image from BigStockPhoto.com

February 21–27

The battles that count aren't the ones for gold medals. The struggles within yourself—the invisible, inevitable battles inside all of us—that's where it's at.

~JESSE OWENS

Date	Qtr.	Sign	Activity
Feb 19, 5:55 am– Feb 21, 1:47 pm	1st	Taurus	Plant annuals for hardiness. Trim to increase growth.
Feb 23, 6:29 pm– Feb 25, 8:08 pm	2nd	Cancer	Plant grains, leafy annuals. Fertilize (chemical). Graft or bud plants. Irrigate. Trim to increase growth.

I s easy maintenance a high priority when you think about gardening? If it is, consider planting more ornamental grasses in your garden. Contrary to popular belief, ornamental grasses are well behaved and remain in neat clumps for years. Grasses provide wonderful background texture and color for blooming plants, and the slightest breeze will set them in motion, which is a delight to both watch and listen to.

_____

_____

_____

February 21
7:42 pm EST

FEBRUARY

S	M	T	W	T	F	S
	1	2	3	4	5	6
7	8	9	10	11	12	13
14	15	16	17	18	19	20
21	22	23	24	25	26	27
28						

March

February 28–March 6

*There is no wealth but life. Life, including all its powers of
love, of joy, and of admiration.*

~JOHN RUSKIN

Date	Qtr.	Sign	Activity
Feb 28, 11:38 am– Mar 1, 7:31 pm	3rd	Virgo	Cultivate, especially medicinal plants. Destroy weeds and pests. Trim to retard growth.
Mar 3, 9:11 pm– Mar 6, 2:36 am	3rd	Scorpio	Plant biennials, perennials, bulbs, and roots. Prune. Irrigate. Fertilize (organic).
Mar 6, 2:36 am– Mar 7, 10:42 am	3rd	Sagittarius	Cultivate. Destroy weeds and pests. Harvest fruits and root crops for food. Trim to retard growth.

Ornamental grasses that can be grown as far north as
Zone 4 include:

- *Panicum* v. 'Heavy Metal' (Zone 4)
- *Carex h.* 'Evergold' (Zone 5)
- *Sisyrinchium bermudianum* (Zone 5)
- *Phalaris* var. 'Picta' (Zone 4)
- *Phalaris a.* 'Strawberries and Cream' (Zone 4)

2009 © Dmitry Sergeev. Image from BigStockPhoto.com

○

February 28
11:38 am EST

MARCH

S	M	T	W	T	F	S
	1	2	3	4	5	6
7	8	9	10	11	12	13
14	15	16	17	18	19	20
21	22	23	24	25	26	27
28	29	30	31			

March 7–13

At the heart of gardening there is a belief in the miraculous.

~MIRABEL OSLER

Date	Qtr.	Sign	Activity
Mar 6, 2:36 am– Mar 7, 10:42 am	3rd	Sagittarius	Cultivate. Destroy weeds and pests. Harvest fruits and root crops for food. Trim to retard growth.
Mar 7, 10:42 am– Mar 8, 12:13 pm	4th	Sagittarius	Cultivate. Destroy weeds and pests. Harvest fruits and root crops for food. Trim to retard growth.
Mar 8, 12:13 pm– Mar 11, 12:42 am	4th	Capricorn	Plant potatoes and tubers. Trim to retard growth.
Mar 11, 12:42 am– Mar 13, 1:44 pm	4th	Aquarius	Cultivate. Destroy weeds and pests. Harvest fruits and root crops for food. Trim to retard growth.
Mar 13, 1:44 pm– Mar 15, 5:01 pm	4th	Pisces	Plant biennials, perennials, bulbs, and roots. Prune. Irrigate. Fertilize (organic).

Think ahead if you have unwelcome rabbits in your yard and plan to plant lavender (*Lavendula*) and catnip (*Nepeta*) in areas where you want rabbits to stay away. They don't like these highly aromatic herbs, nor do they like the smell of cats that come around to eat the catnip, or "catmint."

March 7
10:42 am EST

			MARCH			
S	M	T	W	T	F	S
	1	2	3	4	5	6
7	8	9	10	11	12	13
14	15	16	17	18	19	20
21	22	23	24	25	26	27
28	29	30	31			

March 14–20

Seed catalogs are responsible for more unfulfilled fantasies
than Enron and Penthouse combined.

~MICHAEL PERRY, *TRUCK: A LOVE STORY*

Date	Qtr.	Sign	Activity
Mar 13, 1:44 pm– Mar 15, 5:01 pm	4th	Pisces	Plant biennials, perennials, bulbs, and roots. Prune. Irrigate. Fertilize (organic).
Mar 15, 5:01 pm– Mar 16, 2:32 am	1st	Pisces	Plant grains, leafy annuals. Fertilize (chemical). Graft or bud plants. Irrigate. Trim to increase growth.
Mar 18, 12:29 pm– Mar 20, 8:28 pm	1st	Taurus	Plant annuals for hardiness. Trim to increase growth.

Everyone learns that the Sun rises in the east and sets in the west, but that's actually true only two days out of the year—on the equinoxes. The other 363 days of the year, the Sun rises either north or south of east, and it sets either north or south of west. This year, the Spring Equinox is on March 20, and the Fall Equinox is on September 23.

2009 © Mona Makela. Image from BigStockPhoto.com

March 15
5:01 pm EDT
Daylight Saving Time begins
March 14, 2:00 am

		MARCH				
S	M	T	W	T	F	S
	1	2	3	4	5	6
7	8	9	10	11	12	13
14	15	16	17	18	19	20
21	22	23	24	25	26	27
28	29	30	31			

March 21–27

Sweet spring, full of sweet days and roses.

~GEORGE HERBERT

Date	Qtr.	Sign	Activity
Mar 23, 2:16 am– Mar 23, 7:00 am	1st	Cancer	Plant grains, leafy annuals. Fertilize (chemical). Graft or bud plants. Irrigate. Trim to increase growth.
Mar 23, 7:00 am– Mar 25, 5:39 am	2nd	Cancer	Plant grains, leafy annuals. Fertilize (chemical). Graft or bud plants. Irrigate. Trim to increase growth.

Bells of Ireland, a member of the mint family, are not from Ireland as their name implies. They are native to western Asia, Turkey, and Syria. To have plants ready for the garden in eight to ten weeks, sow seeds indoors now. Germination can take up to a month, but storing the seeds in a plastic bag in your refrigerator for a couple of weeks first will speed up the process. Wait until nighttime temperatures are up to 40°F before planting outdoors in a sunny to partially sunny spot.

_____

_____

_____

_____

March 23
7:00 am EDT

MARCH

S	M	T	W	T	F	S
		1	2	3	4	5	6
7	8	9	10	11	12	13
14	15	16	17	18	19	20
21	22	23	24	25	26	27
28	29	30	31			

2009 © Hazel Proudlove. Image from BigStockPhoto.com

April

March 28–April 3

A friend may well be reckoned the masterpiece of Nature.
~Ralph Waldo Emerson

Date	Qtr.	Sign	Activity
Mar 29, 7:21 am– Mar 29, 10:25 pm	2nd	Libra	Plant annuals for fragrance and beauty. Trim to increase growth.
Mar 31, 8:41 am– Apr 2, 12:53 pm	3rd	Scorpio	Plant biennials, perennials, bulbs, and roots. Prune. Irrigate. Fertilize (organic).
Apr 2, 12:53 pm– Apr 4, 9:07 pm	3rd	Sagittarius	Cultivate. Destroy weeds and pests. Harvest fruits and root crops for food. Trim to retard growth.

Do you know that woodpeckers can peck up to sixteen times per second? Drumming is the term given to the sound of rapid pecking on wood or even metal. Woodpeckers drum for two reasons: to stake out their territory and to find insects. Their springtime drumming is usually done to mark territory, and it can be more annoying to hear than it is damaging to buildings.

March 29
10:25 pm EDT

			April			
S	M	T	W	T	F	S
				1	2	3
4	5	6	7	8	9	10
11	12	13	14	15	16	17
18	19	20	21	22	23	24
25	26	27	28	29	30	

 **April 4–10**

Our life is frittered away by detail . . . Simplify, simplify.

~HENRY DAVID THOREAU

Date	Qtr.	Sign	Activity
Apr 2, 12:53 pm– Apr 4, 9:07 pm	3rd	Sagittarius	Cultivate. Destroy weeds and pests. Harvest fruits and root crops for food. Trim to retard growth.
Apr 4, 9:07 pm– Apr 6, 5:37 am	3rd	Capricorn	Plant potatoes and tubers. Trim to retard growth.
Apr 6, 5:37 am– Apr 7, 8:51 am	4th	Capricorn	Plant potatoes and tubers. Trim to retard growth.
Apr 7, 8:51 am– Apr 9, 9:48 pm	4th	Aquarius	Cultivate. Destroy weeds and pests. Harvest fruits and root crops for food. Trim to retard growth.
Apr 9, 9:48 pm– Apr 12, 9:31 am	4th	Pisces	Plant biennials, perennials, bulbs, and roots. Prune. Irrigate. Fertilize (organic).

Simplify! It'll save you time; it may even add years to your life. Here's a short list of things to do so you can have more time and less work: Identify what is really important to you and eliminate the rest (less to clean). Learn to say no more often (more time for yourself). Limit your buying habits (you won't have to work longer hours to pay for more stuff). Eat more slowly (you'll lose weight and have less indigestion).

April 6
5:37 am EDT

APRIL

S	M	T	W	T	F	S
				1	2	3
4	5	6	7	8	9	10
11	12	13	14	15	16	17
18	19	20	21	22	23	24
25	26	27	28	29	30	

April 11–17 ♈

I love deadlines. I like the whooshing sound they make as they fly by.

~Douglas Adams

Date	Qtr.	Sign	Activity
Apr 9, 9:48 pm–Apr 12, 9:31 am	4th	Pisces	Plant biennials, perennials, bulbs, and roots. Prune. Irrigate. Fertilize (organic).
Apr 12, 9:31 am–Apr 14, 8:29 am	4th	Aries	Cultivate. Destroy weeds and pests. Harvest fruits and root crops for food. Trim to retard growth.
Apr 14, 6:55 pm–Apr 17, 2:08 am	1st	Taurus	Plant annuals for hardiness. Trim to increase growth.

In northern tier states, it's time to prune the apple trees, rake lawns to clean up debris, remove mulch from around the perennials, cut back Russian sage and butterfly bush, and check shrubs and trees for winter damage. Almost everywhere, it's time to sharpen the lawn-mower blade and to begin planting summer bulbs—caladium and gladiolus—at two-week intervals.

April 14
8:29 am EDT

APRIL

S	M	T	W	T	F	S
				1	2	3
4	5	6	7	8	9	10
11	12	13	14	15	16	17
18	19	20	21	22	23	24
25	26	27	28	29	30	

April 18–24

Every child is in a way a genius; and every genius is in a way a child.

~ARTHUR SCHOPENHAUER

Date	Qtr.	Sign	Activity
Apr 19, 7:39 am– Apr 21, 11:42 am	1st	Cancer	Plant grains, leafy annuals. Fertilize (chemical). Graft or bud plants. Irrigate. Trim to increase growth.

Grass seeds do best when planted into freshly prepared soil, but less ideal conditions frequently exist. Good seed/soil contact is essential to getting grass started. Rake the soil to loosen it before broadcasting the seed if you want to add new grass to an existing lawn; using more seed will improve germination.

April 21
2:20 pm EDT

APRIL

S	M	T	W	T	F	S
				1	2	3
4	5	6	7	8	9	10
11	12	13	14	15	16	17
18	19	20	21	22	23	24
25	26	27	28	29	30	

2009 © Alina Goncharova. Image from BigStockPhoto.com

May

April 25–May 1

I am fonder of my garden for the trouble it gives me.

~REGINALD FARRER

Date	Qtr.	Sign	Activity
Apr 25, 4:16 pm–Apr 27, 6:28 pm	2nd	Libra	Plant annuals for fragrance and beauty. Trim to increase growth.
Apr 27, 6:28 pm–Apr 28, 8:19 am	2nd	Scorpio	Plant grains, leafy annuals. Fertilize (chemical). Graft or bud plants. Irrigate. Trim to increase growth.
Apr 28, 8:19 am–Apr 29, 10:36 pm	3rd	Scorpio	Plant biennials, perennials, bulbs, and roots. Prune. Irrigate. Fertilize (organic).
Apr 29, 10:36 pm–May 2, 6:00 am	3rd	Sagittarius	Cultivate. Destroy weeds and pests. Harvest fruits and root crops for food. Trim to retard growth.

In places like Russia, China, and Korea, May Day was recognized as a working-class holiday, similar to Labor Day in the United States. May Day, or International Workers' Day, was born out of the struggle for an eight-hour workday, but the movement became associated with communism and socialism in the late 1800s and quickly fell out of favor in the United States.

2009 © Kati Molin. Image from BigStockPhoto.com

April 28
8:19 am EDT

MAY

S	M	T	W	T	F	S
						1
2	3	4	5	6	7	8
9	10	11	12	13	14	15
16	17	18	19	20	21	22
23	24	25	26	27	28	29
30	31					

May 2–8

My heart leaps when I behold
A rainbow in the sky.

~WILLIAM WORDSWORTH

Date	Qtr.	Sign	Activity
Apr 29, 10:36 pm–May 2, 6:00 am	3rd	Sagittarius	Cultivate. Destroy weeds and pests. Harvest fruits and root crops for food. Trim to retard growth.
May 2, 6:00 am–May 4, 4:52 pm	3rd	Capricorn	Plant potatoes and tubers. Trim to retard growth.
May 4, 4:52 pm–May 6, 12:15 am	3rd	Aquarius	Cultivate. Destroy weeds and pests. Harvest fruits and root crops for food. Trim to retard growth.
May 6, 12:15 am–May 7, 5:34 am	4th	Aquarius	Cultivate. Destroy weeds and pests. Harvest fruits and root crops for food. Trim to retard growth.
May 7, 5:34 am–May 9, 5:29 pm	4th	Pisces	Plant biennials, perennials, bulbs, and roots. Prune. Irrigate. Fertilize (organic).

Potatoes should be stored in a dark, dry, cool place, but not in the refrigerator. Very cool temperatures cause the starch in potatoes to turn into sugar. If you live in a warm climate and must store your potatoes in the refrigerator to keep them from sprouting, let them sit at room temperature for a few hours before you cook them, and the sugar will return to starch.

May 6
12:15 am EDT

MAY

S	M	T	W	T	F	S
						1
2	3	4	5	6	7	8
9	10	11	12	13	14	15
16	17	18	19	20	21	22
23	24	25	26	27	28	29
30	31					

May 9–15

All I am I owe to my mother.

~GEORGE WASHINGTON

Date	Qtr.	Sign	Activity
May 7, 5:34 am–May 9, 5:29 pm	4th	Pisces	Plant biennials, perennials, bulbs, and roots. Prune. Irrigate. Fertilize (organic).
May 9, 5:29 pm–May 12, 2:48 am	4th	Aries	Cultivate. Destroy weeds and pests. Harvest fruits and root crops for food. Trim to retard growth.
May 12, 2:48 am–May 13, 9:04 pm	4th	Taurus	Plant potatoes and tubers. Trim to retard growth.
May 13, 9:04 pm–May 14, 9:18 am	1st	Taurus	Plant annuals for hardiness. Trim to increase growth.

All blue potatoes and 'Delta Blues' (purple potatoes) continue to be popular in U.S. specialty markets, where manufacturers turn them into colorful chips. Purple potatoes are gaining popularity among consumers, too, for mashed or roasted potatoes and in salads. They lose a little of their vibrancy when cooked, but they hold their shape and color well. They also contain the same powerful antioxident found in blueberries.

2009 © Pavel Losevsky. Image from BigStockPhoto.com

● *May 13*
9:04 pm EDT

MAY

S	M	T	W	T	F	S
						1
2	3	4	5	6	7	8
9	10	11	12	13	14	15
16	17	18	19	20	21	22
23	24	25	26	27	28	29
30	31					

May 16–22

The emotional appeal of nature is tremendous, sometimes almost more than one can bear.

~Jan Smuts

Date	Qtr.	Sign	Activity
May 16, 1:46 pm– May 18, 5:06 pm	1st	Cancer	Plant grains, leafy annuals. Fertilize (chemical). Graft or bud plants. Irrigate. Trim to increase growth.
May 22, 10:50 pm– May 25, 2:17 am	2nd	Libra	Plant annuals for fragrance and beauty. Trim to increase growth.

Neither paving stones nor fence posts should be set when the Moon is waxing in Cancer. Cancer days are unfavorable because stones and fence posts will loosen by themselves or rot more quickly. Choose a day, instead, when the Moon is waning and in an earth sign—Taurus, Virgo, or Capricorn.

_____

_____

_____

_____

May 20
7:43 pm EDT

MAY
S	M	T	W	T	F	S
						1
2	3	4	5	6	7	8
9	10	11	12	13	14	15
16	17	18	19	20	21	22
23	24	25	26	27	28	29
30	31					

2009 © John Bunch. Image from BigStockPhoto.com

May 23–29

We need wilderness whether or not we ever set foot in it. We need a refuge even though we may not ever need to go there.

*~*Edward Abbey

Date	Qtr.	Sign	Activity
May 22, 10:50 pm–May 25, 2:17 am	2nd	Libra	Plant annuals for fragrance and beauty. Trim to increase growth.
May 25, 2:17 am–May 27, 7:15 am	2nd	Scorpio	Plant grains, leafy annuals. Fertilize (chemical). Graft or bud plants. Irrigate. Trim to increase growth.
May 27, 7:07 pm–May 29, 2:44 pm	3rd	Sagittarius	Cultivate. Destroy weeds and pests. Harvest fruits and root crops for food. Trim to retard growth.
May 29, 2:44 pm–Jun 1, 1:08 am	3rd	Capricorn	Plant potatoes and tubers. Trim to retard growth.

Fast facts about U.S. wilderness areas:

- Rocks and Island Wilderness in California is the smallest with a total of 5 acres. Wrangell-Saint Elias Wilderness in Alaska is the largest with a total of 9,078,675 acres.
- Wild Sky Wilderness in Washington became the newest wilderness area in May 2008.
- Recreational use of wilderness areas has increased tenfold in forty years; overuse is the biggest threat to the parks.

2009 © Alan Heartfield. Image from BigStockPhoto.com

○

May 27
7:07 pm EDT

MAY

S	M	T	W	T	F	S
						1
2	3	4	5	6	7	8
9	10	11	12	13	14	15
16	17	18	19	20	21	22
23	24	25	26	27	28	29
30	31					

June

May 30–June 5

A gardener learns more in the mistakes than in the successes.

~Barbara Dodge Borland

Date	Qtr.	Sign	Activity
May 29, 2:44 pm–Jun 1, 1:08 am	3rd	Capricorn	Plant potatoes and tubers. Trim to retard growth.
Jun 1, 1:08 am–Jun 3, 1:34 pm	3rd	Aquarius	Cultivate. Destroy weeds and pests. Harvest fruits and root crops for food. Trim to retard growth.
Jun 3, 1:34 pm–Jun 4, 6:13 pm	3rd	Pisces	Plant biennials, perennials, bulbs, and roots. Prune. Irrigate. Fertilize (organic).
Jun 4, 6:13 pm–Jun 6, 1:50 am	4th	Pisces	Plant biennials, perennials, bulbs, and roots. Prune. Irrigate. Fertilize (organic).

Lilacs came to the United States in the 1600s and, today, Rochester, New York, is the lilac capital of the world. Highland Park in Rochester is the home of over 500 varieties of lilacs (*Syringa vulgaris*) and more than 1,200 bushes.

_____

_____

_____

◑

June 4
6:13 pm EDT

		JUNE				
S	M	T	W	T	F	S
		1	2	3	4	5
6	7	8	9	10	11	12
13	14	15	16	17	18	19
20	21	22	23	24	25	26
27	28	29	30			

2009 © Oleg Mitiukhin. Image from BigStockPhoto.com

June 6–12

I am in love with the green earth.

~CHARLES LAMB

Date	Qtr.	Sign	Activity
Jun 4, 6:13 pm– Jun 6, 1:50 am	4th	Pisces	Plant biennials, perennials, bulbs, and roots. Prune. Irrigate. Fertilize (organic).
Jun 6, 1:50 am– Jun 8, 11:41 am	4th	Aries	Cultivate. Destroy weeds and pests. Harvest fruits and root crops for food. Trim to retard growth.
Jun 8, 11:41 am– Jun 10, 6:11 pm	4th	Taurus	Plant potatoes and tubers. Trim to retard growth.
Jun 10, 6:11 pm– Jun 12, 7:15 am	4th	Gemini	Cultivate. Destroy weeds and pests. Harvest fruits and root crops for food. Trim to retard growth.
Jun 12, 9:50 pm– Jun 14, 11:54 pm	1st	Cancer	Plant grains, leafy annuals. Fertilize (chemical). Graft or bud plants. Irrigate. Trim to increase growth.

The time to prune lilacs is right after they're done blooming. Remove the spent blooms, but don't trim off too much or you'll remove next year's flowers. Trim larger stems from the center of dense plants to encourage air circulation and limit plant disease.

_____

_____

_____

_____

June 12
7:15 am EDT

JUNE

S	M	T	W	T	F	S
	1	2	3	4	5	
6	7	8	9	10	11	12
13	14	15	16	17	18	19
20	21	22	23	24	25	26
27	28	29	30			

June 13–19

What's so amazing that keeps us stargazing?
And what do we think we might see?

~Paul Williams, "The Rainbow Connection"

Date	Qtr.	Sign	Activity
Jun 12, 9:50 pm– Jun 14, 11:54 pm	1st	Cancer	Plant grains, leafy annuals. Fertilize (chemical). Graft or bud plants. Irrigate. Trim to increase growth.
Jun 19, 4:13 am– Jun 21, 8:14 am	2nd	Libra	Plant annuals for fragrance and beauty. Trim to increase growth.

Ever wonder what makes a flower blue? It's the pigment del-phinidin, which is found in delphiniums, some petunias, and bachelor buttons, but not in roses. A dark-burgundy rose was produced in 2004 after delphinidin from a petunia was crossed with a mauve-colored rose, but so far growers have been unsuccessful in producing a truly blue rose.

June 19
12:30 am EDT

		June				
S	M	T	W	T	F	S
		1	2	3	4	5
6	7	8	9	10	11	12
13	14	15	16	17	18	19
20	21	22	23	24	25	26
27	28	29	30			

June 20–26

Gardens are a form of autobiography.

~SYDNEY EDDISON

Date	Qtr.	Sign	Activity
Jun 19, 4:13 am– Jun 21, 8:14 am	2nd	Libra	Plant annuals for fragrance and beauty. Trim to increase growth.
Jun 21, 8:14 am– Jun 23, 2:10 pm	2nd	Scorpio	Plant grains, leafy annuals. Fertilize (chemical). Graft or bud plants. Irrigate. Trim to increase growth.
Jun 25, 10:21 pm– Jun 26, 7:30 am	2nd	Capricorn	Graft or bud plants. Trim to increase growth.
Jun 26, 7:30 am– Jun 28, 8:52 am	3rd	Capricorn	Plant potatoes and tubers. Trim to retard growth.

Food, water, and shelter are the three elements that will attract birds to your backyard. If you provide food and water for birds during the summer and fall, plan to do so during the winter, as well. The sound of moving water attracts them during the summer, and open water is precious to birds during the winter.

2009 © Lynne Menturweck

June 26
7:30 am EDT

JUNE

S	M	T	W	T	F	S
		1	2	3	4	5
6	7	8	9	10	11	12
13	14	15	16	17	18	19
20	21	22	23	24	25	26
27	28	29	30			

July

June 27–July 3

*Life can only be understood backwards; but it must be
lived forwards.*

~Søren Kierkegaard

Date	Qtr.	Sign	Activity
Jun 26, 7:30 am– Jun 28, 8:52 am	3rd	Capricorn	Plant potatoes and tubers. Trim to retard growth.
Jun 28, 8:52 am– Jun 30, 9:10 pm	3rd	Aquarius	Cultivate. Destroy weeds and pests. Harvest fruits and root crops for food. Trim to retard growth.
Jun 30, 9:10 pm– Jul 3, 9:44 am	3rd	Pisces	Plant biennials, perennials, bulbs, and roots. Prune. Irrigate. Fertilize (organic).
Jul 3, 9:44 am– Jul 4, 10:35 am	3rd	Aries	Cultivate. Destroy weeds and pests. Harvest fruits and root crops for food. Trim to retard growth.

In the 1690s, the word *picnic* was used to describe a group of people dining in a restaurant who brought their own wine. Today's picnics are often associated with holidays, family outings, and romance, but getting caught in rush hour traffic or in the rain is "no picnic."

JULY

S	M	T	W	T	F	S
				1	2	3
4	5	6	7	8	9	10
11	12	13	14	15	16	17
18	19	20	21	22	23	24
25	26	27	28	29	30	31

2009 © Rain Roogla. Image from BigStockPhoto.com

July 4–10

Once you learn to read, you will be forever free.

~FREDERICK DOUGLASS

Date	Qtr.	Sign	Activity
Jul 3, 9:44 am– Jul 4, 10:35 am	3rd	Aries	Cultivate. Destroy weeds and pests. Harvest fruits and root crops for food. Trim to retard growth.
Jul 4, 10:35 am– Jul 5, 8:29 pm	4th	Aries	Cultivate. Destroy weeds and pests. Harvest fruits and root crops for food. Trim to retard growth.
Jul 5, 8:29 pm– Jul 8, 3:51 am	4th	Taurus	Plant potatoes and tubers. Trim to retard growth.
Jul 8, 3:51 am– Jul 10, 7:38 am	4th	Gemini	Cultivate. Destroy weeds and pests. Harvest fruits and root crops for food. Trim to retard growth.
Jul 10, 7:38 am– Jul 11, 3:40 pm	4th	Cancer	Plant biennials, perennials, bulbs, and roots. Prune. Irrigate. Fertilize (organic).

Keep your picnics safe by following FDA guidelines for storing food in warm weather. Cold food needs to be stored below 40°F. Pack securely wrapped meat, poultry, and seafood while still frozen, and consider packing beverages in one cooler and perishables in another. Keep coolers closed as much as possible.

2009 © Daniel Sroga. Image from BigStockPhoto.com

◑

July 4
10:35 am EDT

JULY

S	M	T	W	T	F	S
				1	2	3
4	5	6	7	8	9	10
11	12	13	14	15	16	17
18	19	20	21	22	23	24
25	26	27	28	29	30	31

July 11–17

*Give me the splendid silent sun with all his beams
full-dazzling.*

~WALT WHITMAN

Date	Qtr.	Sign	Activity
Jul 10, 7:38 am–Jul 11, 3:40 pm	4th	Cancer	Plant biennials, perennials, bulbs, and roots. Prune. Irrigate. Fertilize (organic).
Jul 11, 3:40 pm–Jul 12, 8:53 am	1st	Cancer	Plant grains, leafy annuals. Fertilize (chemical). Graft or bud plants. Irrigate. Trim to increase growth.
Jul 16, 10:24 am–Jul 18, 6:11 am	1st	Libra	Plant annuals for fragrance and beauty. Trim to increase growth.

Hot foods should be kept hot—at or above 140°F—even in the summer. Wrap cooked food and place it in an insulated container until it's served. Warm foods should not sit out more than two hours (one hour in temperatures above 90°F). If food sits out more than two hours, throw it away.

_____

_____

_____

_____

●
July 11
3:40 pm EDT

JULY

S	M	T	W	T	F	S
				1	2	3
4	5	6	7	8	9	10
11	12	13	14	15	16	17
18	19	20	21	22	23	24
25	26	27	28	29	30	31

July 18–24

Those who don't believe in magic will never find it.

~ROALD DAHL

Date	Qtr.	Sign	Activity
Jul 16, 10:24 am– Jul 18, 6:11 am	1st	Libra	Plant annuals for fragrance and beauty. Trim to increase growth.
Jul 18, 6:11 am– Jul 18, 1:42 pm	2nd	Libra	Plant annuals for fragrance and beauty. Trim to increase growth.
Jul 18, 1:42 pm– Jul 20, 7:48 pm	2nd	Scorpio	Plant grains, leafy annuals. Fertilize (chemical). Graft or bud plants. Irrigate. Trim to increase growth.
Jul 23, 4:39 am– Jul 25, 3:38 pm	2nd	Capricorn	Graft or bud plants. Trim to increase growth.

The *Syringa pekinensis* (Peking lilac) blooms in July and can be grown in Zone 4. The genus name for lilacs (*Syringa*) is derived from the Greek work *syrinx* meaning "hollow stem." Ancient Greek doctors supposedly used these stems to inject medications into their patients.

2009 © Philip Lange. Image from BigStockPhoto.com

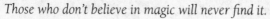

July 18
6:11 am EDT

JULY

S	M	T	W	T	F	S
				1	2	3
4	5	6	7	8	9	10
11	12	13	14	15	16	17
18	19	20	21	22	23	24
25	26	27	28	29	30	31

July 25–31

Green fingers are the extension of a verdant heart.

~RUSSELL PAGE

Date	Qtr.	Sign	Activity
Jul 23, 4:39 am– Jul 25, 3:38 pm	2nd	Capricorn	Graft or bud plants. Trim to increase growth.
Jul 25, 9:37 pm– Jul 28, 4:00 am	3rd	Aquarius	Cultivate. Destroy weeds and pests. Harvest fruits and root crops for food. Trim to retard growth.
Jul 28, 4:00 am– Jul 30, 4:42 pm	3rd	Pisces	Plant biennials, perennials, bulbs, and roots. Prune. Irrigate. Fertilize (organic).
Jul 30, 4:42 pm– Aug 2, 4:13 am	3rd	Aries	Cultivate. Destroy weeds and pests. Harvest fruits and root crops for food. Trim to retard growth.

The quality of air inside our homes can be very poor at times, especially when we keep windows and doors closed against the heat. Air conditioners only recirculate the air. Keeping houseplants can add a nice touch of green or solve a decorating problem, but plants are also good air cleaners. Aloe vera, bamboo palm, green spider plant, English or common ivy, elephant ear philodendron, golden pathos, and peace lily are among the most effective at filtering pollution from the air.

July 25
9:37 pm EDT

JULY

S	M	T	W	T	F	S
				1	2	3
4	5	6	7	8	9	10
11	12	13	14	15	16	17
18	19	20	21	22	23	24
25	26	27	28	29	30	31

August ♌

August 1–7

All that is good is simple and all that is simple is good.
~Mikhail Kalashnikov

Date	Qtr.	Sign	Activity
Jul 30, 4:42 pm– Aug 2, 4:13 am	3rd	Aries	Cultivate. Destroy weeds and pests. Harvest fruits and root crops for food. Trim to retard growth.
Aug 2, 4:13 am– Aug 3, 12:59 am	3rd	Taurus	Plant potatoes and tubers. Trim to retard growth.
Aug 3, 12:59 am– Aug 4, 12:54 pm	4th	Taurus	Plant potatoes and tubers. Trim to retard growth.
Aug 4, 12:54 pm– Aug 6, 5:50 pm	4th	Gemini	Cultivate. Destroy weeds and pests. Harvest fruits and root crops for food. Trim to retard growth.
Aug 6, 5:50 pm– Aug 8, 7:23 pm	4th	Cancer	Plant biennials, perennials, bulbs, and roots. Prune. Irrigate. Fertilize (organic).

If there's a brown cloud hanging over the city you live in this summer, you can make a difference by adopting more "green living" habits. Walk more or ride a bike. If that's not practical, plan ahead to reduce the number of trips made in your car.

August 3
12:59 am EDT

August

S	M	T	W	T	F	S
						1
2	3	4	5	6	7	8
9	10	11	12	13	14	15
16	17	18	19	20	21	22
23	24	25	26	27	28	29
30	31					

August 8–14

Hope is a good breakfast, but it is a bad supper.

~Francis Bacon

Date	Qtr.	Sign	Activity
Aug 6, 5:50 pm– Aug 8, 7:23 pm	4th	Cancer	Plant biennials, perennials, bulbs, and roots. Prune. Irrigate. Fertilize (organic).
Aug 8, 7:23 pm– Aug 9, 11:08 pm	4th	Leo	Cultivate. Destroy weeds and pests. Harvest fruits and root crops for food. Trim to retard growth.
Aug 12, 6:43 pm– Aug 14, 8:26 pm	1st	Libra	Plant annuals for fragrance and beauty. Trim to increase growth.
Aug 14, 8:26 pm– Aug 16, 2:14 pm	1st	Scorpio	Plant grains, leafy annuals. Fertilize (chemical). Graft or bud plants. Irrigate. Trim to increase growth.

Some root crops are bunched and sold with their tops intact. While vegetables with green tops may look pretty, their storage life is shorter than for those that have been topped. For example, bunched beets can be stored for up to two weeks, while beets without tops can be stored for several months.

August 9
11:08 pm EDT

August

S	M	T	W	T	F	S
					1	
1	2	3	4	5	6	7
8	9	10	11	12	13	14
15	16	17	18	19	20	21
22	23	24	25	26	27	28
29	30	31				

August 15–21

And life is colour and warmth and light
And a striving evermore for these.

~JULIAN GRENFELL

Date	Qtr.	Sign	Activity
Aug 14, 8:26 pm–Aug 16, 2:14 pm	1st	Scorpio	Plant grains, leafy annuals. Fertilize (chemical). Graft or bud plants. Irrigate. Trim to increase growth.
Aug 16, 2:14 pm–Aug 17, 1:34 am	2nd	Scorpio	Plant grains, leafy annuals. Fertilize (chemical). Graft or bud plants. Irrigate. Trim to increase growth.
Aug 19, 10:17 am–Aug 21, 9:37 pm	2nd	Capricorn	Graft or bud plants. Trim to increase growth.

If you were a color, what color would you be? You can find out by taking the "What color are you?" quiz at http://www.spacefem.com that will match your personality to one of 144 colors. It's fun and easy to do. Do you doubt the results? The "What color are you?" quiz at http://www.quizilla.com will give you a second opinion.

August 16
2:14 pm EDT

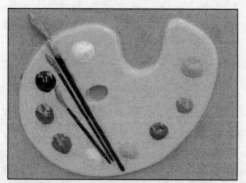

2009 © Linda Brumbaugh. Image from BigStockPhoto.com

AUGUST

S	M	T	W	T	F	S
1	2	3	4	5	6	7
8	9	10	11	12	13	14
15	16	17	18	19	20	21
22	23	24	25	26	27	28
29	30	31				

♍ August 22–28

*Life's meaning has always eluded me and I guess it always
will. But I love it just the same.*

~E. B. WHITE

Date	Qtr.	Sign	Activity
Aug 24, 10:11 am– Aug 24, 1:05 pm	2nd	Pisces	Plant grains, leafy annuals. Fertilize (chemical). Graft or bud plants. Irrigate. Trim to increase growth.
Aug 24, 1:05 pm– Aug 26, 10:49 pm	3rd	Pisces	Plant biennials, perennials, bulbs, and roots. Prune. Irrigate. Fertilize (organic).
Aug 26, 10:49 pm– Aug 29, 10:35 am	3rd	Aries	Cultivate. Destroy weeds and pests. Harvest fruits and root crops for food. Trim to retard growth.

According to Michael Ober, master cartwright in St. Johann, Tyrol, and recorded by Josef Schmutzer in December 1912, timber for sawing should be cut down when the Moon is waxing in Pisces; then boards and timber will not be worm-ridden.

_____

_____

_____

_____

○
August 24
1:05 pm EDT

AUGUST

S	M	T	W	T	F	S
1	2	3	4	5	6	7
8	9	10	11	12	13	14
15	16	17	18	19	20	21
22	23	24	25	26	27	28
29	30	31				

September ♍

August 29–September 4

The books we love, love us back.

~JOHN LEONARD

Date	Qtr.	Sign	Activity
Aug 26, 10:49 pm–Aug 29, 10:35 am	3rd	Aries	Cultivate. Destroy weeds and pests. Harvest fruits and root crops for food. Trim to retard growth.
Aug 29, 10:35 am–Aug 31, 8:19 pm	3rd	Taurus	Plant potatoes and tubers. Trim to retard growth.
Aug 31, 8:19 pm–Sep 1, 1:22 pm	3rd	Gemini	Cultivate. Destroy weeds and pests. Harvest fruits and root crops for food. Trim to retard growth.
Sep 1, 1:22 pm–Sep 3, 2:50 am	4th	Gemini	Cultivate. Destroy weeds and pests. Harvest fruits and root crops for food. Trim to retard growth.
Sep 3, 2:50 am–Sep 5, 5:45 am	4th	Cancer	Plant biennials, perennials, bulbs, and roots. Prune. Irrigate. Fertilize (organic).

Use this week to buy goods, while the Moon is waning in Gemini, a mutable sign. Merchants want to reduce their inventory, and you'll see more sales after the Full Moon—perfect timing for back-to-school shopping.

September 1
1:22 pm EDT

SEPTEMBER

S	M	T	W	T	F	S
			1	2	3	4
5	6	7	8	9	10	11
12	13	14	15	16	17	18
19	20	21	22	23	24	25
26	27	28	29	30		

♍ September 5–11

True happiness
Consists not in the multitude of friends,
But in the worth and choice.

~Ben Jonson

Date	Qtr.	Sign	Activity
Sep 3, 2:50 am– Sep 5, 5:45 am	4th	Cancer	Plant biennials, perennials, bulbs, and roots. Prune. Irrigate. Fertilize (organic).
Sep 5, 5:45 am– Sep 7, 5:53 am	4th	Leo	Cultivate. Destroy weeds and pests. Harvest fruits and root crops for food. Trim to retard growth.
Sep 7, 5:53 am– Sep 8, 6:30 am	4th	Virgo	Cultivate, especially medicinal plants. Destroy weeds and pests. Trim to retard growth.
Sep 9, 5:01 am– Sep 11, 5:21 am	1st	Libra	Plant annuals for fragrance and beauty. Trim to increase growth.
Sep 11, 5:21 am– Sep 13, 8:52 am	1st	Scorpio	Plant grains, leafy annuals. Fertilize (chemical). Graft or bud plants. Irrigate. Trim to increase growth.

Decorate a room this week and plan the project by the Moon. Make your plans on Monday, shop on Tuesday, take care of details on Wednesday, "load" the room on Thursday, and add final touches on Friday. It'll be perfect!

September 8
6:30 am EDT

	SEPTEMBER					
S	M	T	W	T	F	S
				1	2	3	4
5	6	7	8	9	10	11
12	13	14	15	16	17	18
19	20	21	22	23	24	25
26	27	28	29	30		

September 12–18 ♍

*Don't judge each day by the harvest you reap, but by the
seeds you plant.*

~ROBERT LOUIS STEVENSON

Date	Qtr.	Sign	Activity
Sep 11, 5:21 am– Sep 13, 8:52 am	1st	Scorpio	Plant grains, leafy annuals. Fertilize (chemical). Graft or bud plants. Irrigate. Trim to increase growth.
Sep 15, 4:30 pm– Sep 18, 3:35 am	2nd	Capricorn	Graft or bud plants. Trim to increase growth.

Yellow-jacket wasps are unwelcome intruders at backyard get-togethers this time of year. They are a scavenger species, and they become more aggressive in the fall when their colonies reach maximum size. Trapping wasps is a temporary solution to the problem. The most effective traps use a synthetic attractant called n-heptyl butyrate to lure the wasps into the traps.

_____

_____

_____

September 15
1:50 am EDT

		SEPTEMBER				
S	M	T	W	T	F	S
			1	2	3	4
5	6	7	8	9	10	11
12	13	14	15	16	17	18
19	20	21	22	23	24	25
26	27	28	29	30		

2009 © Alexandr Ozerov. Image from BigStockPhoto.com

♍ September 19–25

Delicious autumn! My very soul is wedded to it, and if I were a bird I would fly about the earth seeking successive autumns.

~GEORGE ELIOT

Date	Qtr.	Sign	Activity
Sep 20, 4:15 pm– Sep 23, 4:47 am	2nd	Pisces	Plant grains, leafy annuals. Fertilize (chemical). Graft or bud plants. Irrigate. Trim to increase growth.
Sep 23, 5:17 am– Sep 25, 4:17 pm	3rd	Aries	Cultivate. Destroy weeds and pests. Harvest fruits and root crops for food. Trim to retard growth.
Sep 25, 4:17 pm– Sep 28, 2:10 am	3rd	Taurus	Plant potatoes and tubers. Trim to retard growth.

Grapes and other fruits hang heavy on the vine this time of year. Early in the week, while the Moon is in Pisces, is a perfect time to begin a batch of wine. Whether the grapes come from a vineyard or your own backyard, you can make an impressive wine in ten to fourteen days, with much of the work being done in the first week. Start now and enjoy your own label over the holidays.

_____

_____

_____

○
September 23
5:17 am EDT

SEPTEMBER

S	M	T	W	T	F	S
			1	2	3	4
5	6	7	8	9	10	11
12	13	14	15	16	17	18
19	20	21	22	23	24	25
26	27	28	29	30		

October

September 26–October 2

It is of immense importance to learn to laugh at ourselves.

~KATHERINE MANSFIELD

Date	Qtr.	Sign	Activity
Sep 25, 4:17 pm– Sep 28, 2:10 am	3rd	Taurus	Plant potatoes and tubers. Trim to retard growth.
Sep 28, 2:10 am– Sep 30, 9:46 am	3rd	Gemini	Cultivate. Destroy weeds and pests. Harvest fruits and root crops for food. Trim to retard growth.
Sep 30, 9:46 am– Sep 30, 11:52 pm	3rd	Cancer	Plant biennials, perennials, bulbs, and roots. Prune. Irrigate. Fertilize (organic).
Sep 30, 11:52 pm– Oct 2, 2:21 pm	4th	Cancer	Plant biennials, perennials, bulbs, and roots. Prune. Irrigate. Fertilize (organic).
Oct 2, 2:21 pm– Oct 4, 4:00 pm	4th	Leo	Cultivate. Destroy weeds and pests. Harvest fruits and root crops for food. Trim to retard growth.

Laughing at yourself is good for your health and it's a lot cheaper than Prozac. Step one for many of us is to give up the need to be perfect. Negative habits are easier to break when the Moon is waning.

2009 © Patrick Moore. Image from BigStockPhoto.com

September 30
11:52 pm EDT

OCTOBER

S	M	T	W	T	F	S
					1	2
3	4	5	6	7	8	9
10	11	12	13	14	15	16
17	18	19	20	21	22	23
24	25	26	27	28	29	30
31						

♎ October 3–9

To hear, one must be silent.

~Ursula K. LeGuin

Date	Qtr.	Sign	Activity
Oct 2, 2:21 pm– Oct 4, 4:00 pm	4th	Leo	Cultivate. Destroy weeds and pests. Harvest fruits and root crops for food. Trim to retard growth.
Oct 4, 4:00 pm– Oct 6, 3:52 pm	4th	Virgo	Cultivate, especially medicinal plants. Destroy weeds and pests. Trim to retard growth.
Oct 7, 2:45 pm– Oct 8, 3:52 pm	1st	Libra	Plant annuals for fragrance and beauty. Trim to increase growth.
Oct 8, 3:52 pm– Oct 10, 6:09 pm	1st	Scorpio	Plant grains, leafy annuals. Fertilize (chemical). Graft or bud plants. Irrigate. Trim to increase growth.

The five best trees to plant in urban areas are:

- Red maple
- Flowering dogwood
- Red and white oak
- American elm
- Redbud

October 7
2:45 pm EDT

OCTOBER

S	M	T	W	T	F	S
					1	2
3	4	5	6	7	8	9
10	11	12	13	14	15	16
17	18	19	20	21	22	23
24	25	26	27	28	29	30
31						

2009 © Vaidas Bucys. Image from BigStockPhoto.com

October 10–16

No seed shall perish which the soul hath sown.

~JOHN ADDINGTON SYMONDS

Date	Qtr.	Sign	Activity
Oct 8, 3:52 pm– Oct 10, 6:09 pm	1st	Scorpio	Plant grains, leafy annuals. Fertilize (chemical). Graft or bud plants. Irrigate. Trim to increase growth.
Oct 13, 12:17 am– Oct 14, 5:27 pm	1st	Capricorn	Graft or bud plants. Trim to increase growth.
Oct 14, 5:27 pm– Oct 15, 10:24 am	2nd	Capricorn	Graft or bud plants. Trim to increase growth.

Tropical plants can survive cold winters if they are properly cared for. Tubers and rhizomes need a cool (40°F to 50°F) place that has good air circulation. Hibiscus, mandevilla, and other tropicals can be brought indoors and allowed to dry out. After the leaves drop off, move the plant to a cool place and water sparingly when the soil is dry, 2 or 3 inches deep.

October 14
5:37 pm EDT

OCTOBER

S	M	T	W	T	F	S
					1	2
3	4	5	6	7	8	9
10	11	12	13	14	15	16
17	18	19	20	21	22	23
24	25	26	27	28	29	30
31						

October 17–23

I don't know the key to success, but the key to failure is trying to please everybody.

~BILL COSBY

Date	Qtr.	Sign	Activity
Oct 17, 10:52 pm– Oct 20, 11:23 am	2nd	Pisces	Plant grains, leafy annuals. Fertilize (chemical). Graft or bud plants. Irrigate. Trim to increase growth.
Oct 22, 9:37 pm– Oct 22, 10:30 pm	3rd	Aries	Cultivate. Destroy weeds and pests. Harvest fruits and root crops for food. Trim to retard growth.
Oct 22, 10:30 pm– Oct 25, 7:47 am	3rd	Taurus	Plant potatoes and tubers. Trim to retard growth.

Ancient people believed that bird flights and calls were omens of future events. Modern scientists watch birds and interpret their activities, too. Changes in bird populations or migrations offer clues about the health of the environment. Some birds, such as the bald eagle, are "indicator" species because they forecast environmental conditions. According to a survey by the U.S. Fish and Wildlife Service, 51.3 million Americans watch birds; it's the fastest-growing outdoor activity in the United States.

○
October 22
9:37 pm EDT

OCTOBER

S	M	T	W	T	F	S
					1	2
3	4	5	6	7	8	9
10	11	12	13	14	15	16
17	18	19	20	21	22	23
24	25	26	27	28	29	30
31						

October 24–30 ♏

Live all you can; it's a mistake not to.

~HENRY JAMES

Date	Qtr.	Sign	Activity
Oct 22, 10:30 pm– Oct 25, 7:47 am	3rd	Taurus	Plant potatoes and tubers. Trim to retard growth.
Oct 25, 7:47 am– Oct 27, 3:14 pm	3rd	Gemini	Cultivate. Destroy weeds and pests. Harvest fruits and root crops for food. Trim to retard growth.
Oct 27, 3:14 pm– Oct 29, 8:39 pm	3rd	Cancer	Plant biennials, perennials, bulbs, and roots. Prune. Irrigate. Fertilize (organic).
Oct 29, 8:39 pm– Oct 30, 8:46 am	3rd	Leo	Cultivate. Destroy weeds and pests. Harvest fruits and root crops for food. Trim to retard growth.
Oct 30, 8:46 am– Oct 31, 11:51 pm	4th	Leo	Cultivate. Destroy weeds and pests. Harvest fruits and root crops for food. Trim to retard growth.

It's time to protect your young trees and shrubs against sunscald, cold damage, and hungry rodents. Sunscald appears as an elongated, cracked area on the south or southwest side of a tree; it can be prevented by using tree wrap or plastic tree guards on the trunk. Tree guards will also protect trees from rodents. Guards should be larger than the trunk and extend 2 to 3 inches below ground level and 18 to 24 inches above the anticipated snow line.

October 30
8:46 am EDT

OCTOBER

S	M	T	W	T	F	S
					1	2
3	4	5	6	7	8	9
10	11	12	13	14	15	16
17	18	19	20	21	22	23
24	25	26	27	28	29	30
31						

♏ November

October 31–November 6

Curiosity may have killed the cat, but it did alright by me.

~HELEN HAYES

Date	Qtr.	Sign	Activity
Oct 30, 8:46 am– Oct 31, 11:51 pm	4th	Leo	Cultivate. Destroy weeds and pests. Harvest fruits and root crops for food. Trim to retard growth.
Oct 31, 11:51 pm– Nov 3, 1:19 am	4th	Virgo	Cultivate, especially medicinal plants. Destroy weeds and pests. Trim to retard growth.
Nov 5, 2:16 am– Nov 6, 12:52 am	4th	Scorpio	Plant biennials, perennials, bulbs, and roots. Prune. Irrigate. Fertilize (organic).
Nov 6, 12:52 am– Nov 7, 3:28 am	1st	Scorpio	Plant grains, leafy annuals. Fertilize (chemical). Graft or bud plants. Irrigate. Trim to increase growth.

Check the gutters on your house. Even if you don't have trees nearby, you may find that they are blocked because gutters seem to magnetically attract debris! Clean them early in the week while the Moon is waning.

_____

_____

_____

November 6
12:52 am EDT

		NOVEMBER				
S	M	T	W	T	F	S
	1	2	3	4	5	6
7	8	9	10	11	12	13
14	15	16	17	18	19	20
21	22	23	24	25	26	27
28	29	30				

2009 © Mitch Aunger. Image from BigStockPhoto.com

November 7–13 ♏

A true friend is someone who is there for you when he'd rather be anywhere else.

—Len Wein

Date	Qtr.	Sign	Activity
Nov 6, 12:52 am–Nov 7, 3:28 am	1st	Scorpio	Plant grains, leafy annuals. Fertilize (chemical). Graft or bud plants. Irrigate. Trim to increase growth.
Nov 9, 8:37 am–Nov 11, 5:32 pm	1st	Capricorn	Graft or bud plants. Trim to increase growth.

Getting rid of rabbits is a high priority for urban residents but there's no easy solution. Rabbits have adapted well to urban living. Humane ways to minimize damage hungry rabbits do to trees, shrubs, and plants include live traps or temporary fences to enclose plantings that are attractive to rabbits. If you set a trap, try baiting it with peanut butter (rabbits like it!) and check the traps often. Release the rabbits you catch in wooded areas away from any backyards.

2009 © Sabrina Ryan. Image from BigStockPhoto.com

◐

November 13
11:39 am EST

Daylight Saving Time ends November 7, 2:00 am

November

S	M	T	W	T	F	S
	1	2	3	4	5	6
7	8	9	10	11	12	13
14	15	16	17	18	19	20
21	22	23	24	25	26	27
28	29	30				

♏ November 14–20

*If the people lived their lives as if it were a song for singing
out of light—provides the music for the stars to be dancing
circles in the night.*

~Hymn to the Russian Earth

Date	Qtr.	Sign	Activity
Nov 14, 5:24 am– Nov 16, 5:59 pm	2nd	Pisces	Plant grains, leafy annuals. Fertilize (chemical). Graft or bud plants. Irrigate. Trim to increase growth.
Nov 19, 5:04 am– Nov 21, 12:27 pm	2nd	Taurus	Plant annuals for hardiness. Trim to increase growth.

Where do houseplant pests come from, and why is the problem more severe during winter? They're on the plants and in the soil, and houseplants are under more stress during the winter so they are more vulnerable to attacks by insects and spiders. Establish a routine for watering and fertilizing your houseplants, and check them regularly for insects. Early treatment with nontoxic insecticide spray or soap can minimize stress or plant loss.

NOVEMBER

S	M	T	W	T	F	S
	1	2	3	4	5	6
7	8	9	10	11	12	13
14	15	16	17	18	19	20
21	22	23	24	25	26	27
28	29	30				

2009 © Cathy Yeulet. Image from BigStockPhoto.com

November 21–27

There is no sincerer love than the love of food.

~GEORGE BERNARD SHAW

Date	Qtr.	Sign	Activity
Nov 19, 5:04 am–Nov 21, 12:27 pm	2nd	Taurus	Plant annuals for hardiness. Trim to increase growth.
Nov 21, 12:27 pm–Nov 21, 1:46 pm	3rd	Taurus	Plant potatoes and tubers. Trim to retard growth.
Nov 21, 1:46 pm–Nov 23, 8:14 pm	3rd	Gemini	Cultivate. Destroy weeds and pests. Harvest fruits and root crops for food. Trim to retard growth.
Nov 23, 8:14 pm–Nov 26, 1:01 am	3rd	Cancer	Plant biennials, perennials, bulbs, and roots. Prune. Irrigate. Fertilize (organic).
Nov 26, 1:01 am–Nov 28, 4:34 am	3rd	Leo	Cultivate. Destroy weeds and pests. Harvest fruits and root crops for food. Trim to retard growth.

African violets, with their furry, soft leaves and delicate blooms, muscle their way into many people's hearts and homes. They're tempting to look at and touch, but many people believe they are also hard to grow. Not so. If you provide them with normal care, they will reward you with almost continuous blooms.

2009 © Arlene Gee. Image from BigStockPhoto.com

November 21
12:27 pm EST

NOVEMBER

S	M	T	W	T	F	S
		1	2	3	4	5	6
7	8	9	10	11	12	13
14	15	16	17	18	19	20
21	22	23	24	25	26	27
28	29	30				

December

November 28–December 4

Nature cannot be ordered about, except by obeying her.

~Francis Bacon

Date	Qtr.	Sign	Activity
Nov 26, 1:01 am– Nov 28, 4:34 am	3rd	Leo	Cultivate. Destroy weeds and pests. Harvest fruits and root crops for food. Trim to retard growth.
Nov 28, 4:34 am– Nov 28, 3:36 pm	3rd	Virgo	Cultivate, especially medicinal plants. Destroy weeds and pests. Trim to retard growth.
Nov 28, 3:36 pm– Nov 30, 7:15 am	4th	Virgo	Cultivate, especially medicinal plants. Destroy weeds and pests. Trim to retard growth.
Dec 2, 9:44 am– Dec 4, 12:59 pm	4th	Scorpio	Plant biennials, perennials, bulbs, and roots. Prune. Irrigate. Fertilize (organic).
Dec 4, 12:59 pm– Dec 5, 12:36 pm	4th	Sagittarius	Cultivate. Destroy weeds and pests. Harvest fruits and root crops for food. Trim to retard growth.

African violets tolerate heat up to about 85°F during the day and cooler nighttime temperatures of about 65°F. They require watering once a week with tepid water and indirect light 8 to 10 hours per day for growth and to promote blooming.

November 28
3:36 pm EST

		DECEMBER				
S	M	T	W	T	F	S
				1	2	3	4
5	6	7	8	9	10	11
12	13	14	15	16	17	18
19	20	21	22	23	24	25
26	27	28	29	30	31	

2009 © Igor Zhorov. Image from BigStockPhoto.com

December 5–11

*Remember that the most beautiful things in the world are the
most useless: peacocks and lilies, for instance.*

~JOHN RUSKIN

Date	Qtr.	Sign	Activity
Dec 4, 12:59 pm– Dec 5, 12:36 pm	4th	Sagittarius	Cultivate. Destroy weeds and pests. Harvest fruits and root crops for food. Trim to retard growth.
Dec 6, 6:16 pm– Dec 9, 2:30 am	1st	Capricorn	Graft or bud plants. Trim to increase growth.
Dec 11, 1:41 pm– Dec 13, 8:59 am	1st	Pisces	Plant grains, leafy annuals. Fertilize (chemical). Graft or bud plants. Irrigate. Trim to increase growth.

I f you want to try a natural face mask, cook carrots in a small
amount of water until tender; then, mash them and apply the
warm paste to your face. Leave on for 15 minutes before rinsing
off. The carrot mask is thought to be good for acne and preventing wrinkles.

_____

_____

_____

_____

2009 © Diego Cervo. Image from BigStockPhoto.com

●

*December 5
12:36 pm EST*

DECEMBER

S	M	T	W	T	F	S
			1	2	3	4
5	6	7	8	9	10	11
12	13	14	15	16	17	18
19	20	21	22	23	24	25
26	27	28	29	30	31	

December 12–18

The thing always happens that you really believe in; and the belief in a thing makes it happen.

~FRANK LLOYD WRIGHT

Date	Qtr.	Sign	Activity
Dec 11, 1:41 pm– Dec 13, 8:59 am	1st	Pisces	Plant grains, leafy annuals. Fertilize (chemical). Graft or bud plants. Irrigate. Trim to increase growth.
Dec 13, 8:59 am– Dec 14, 2:15 am	2nd	Pisces	Plant grains, leafy annuals. Fertilize (chemical). Graft or bud plants. Irrigate. Trim to increase growth.
Dec 16, 1:49 pm– Dec 18, 10:37 pm	2nd	Taurus	Plant annuals for hardiness. Trim to increase growth.

Dreams do come true. Children believe it, and so do many adults. Books have been written about dreams that come true. Martin Luther King, Jr. had a dream, and his dream is coming true. So, in this month of wonder, magic, and merriment, think about a dream you've had—and then believe.

_____

_____

_____

December 13
8:59 am EST

```
         DECEMBER
   S  M  T  W  T  F  S
            1  2  3  4
   5  6  7  8  9 10 11
  12 13 14 15 16 17 18
  19 20 21 22 23 24 25
  26 27 28 29 30 31
```

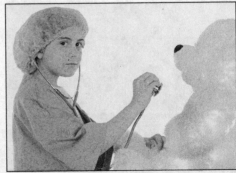

December 19–25

Where there is great love there are always miracles.

~WILLA CATHER

Date	Qtr.	Sign	Activity
Dec 21, 3:13 am– Dec 21, 4:22 am	3rd	Gemini	Cultivate. Destroy weeds and pests. Harvest fruits and root crops for food. Trim to retard growth.
Dec 21, 4:22 am– Dec 23, 7:51 am	3rd	Cancer	Plant biennials, perennials, bulbs, and roots. Prune. Irrigate. Fertilize (organic).
Dec 23, 7:51 am– Dec 25, 10:14 am	3rd	Leo	Cultivate. Destroy weeds and pests. Harvest fruits and root crops for food. Trim to retard growth.
Dec 25, 10:14 am– Dec 27, 12:38 pm	3rd	Virgo	Cultivate, especially medicinal plants. Destroy weeds and pests. Trim to retard growth.

Winter Solstice is celebrated in many Northern Hemisphere countries. Romans called the shortest day of the year the Birthday of the Unconquered Sun, some Scandinavian countries call it Yule Day, and among the Zuni and Hopi tribes, it is Soyal. Sacred ritual chambers, called kivas, are ritually opened on Soyal to mark the beginning of the Kachina season.

○
December 21
3:13 am EST

DECEMBER

S	M	T	W	T	F	S
			1	2	3	4
5	6	7	8	9	10	11
12	13	14	15	16	17	18
19	20	21	22	23	24	25
26	27	28	29	30	31	

December 26–January 1

Be who you are and say what you feel, because those who mind don't matter and those who matter don't mind.

~THEODOR SEUSS GEISEL, AKA DR. SEUSS

Date	Qtr.	Sign	Activity
Dec 25, 10:14 am– Dec 27, 12:38 pm	3rd	Virgo	Cultivate, especially medicinal plants. Destroy weeds and pests. Trim to retard growth.
Dec 29, 3:49 pm– Dec 31, 8:21 pm	4th	Scorpio	Plant biennials, perennials, bulbs, and roots. Prune. Irrigate. Fertilize (organic).
Dec 31, 8:21 pm– Jan 3, 2:39 am	4th	Saggitarius	Cultivate. Destroy weeds and pests. Harvest fruits and root crops for food. Trim to retard growth.

The seeds for cool-season grasses (perennial rye grasses, blue-grasses, and fescue) can be broadcast on the snow in the spring. The seeds are carried down by melting snow and have a chance to start growing before the weather gets warm or the birds carry away all the seeds. Germination is reduced, though, so use more seed to improve the chances of getting grass cover by summer.

_____

_____

_____

December 27
11:18 pm EST

DECEMBER

S	M	T	W	T	F	S
			1	2	3	4
5	6	7	8	9	10	11
12	13	14	15	16	17	18
19	20	21	22	23	24	25
26	27	28	29	30	31	

2009 © Pavel Dunyushkin. Image from BigStockPhoto.com

Gardening by the Moon

Today, people often reject the notion of gardening according to the Moon's phase and sign. The usual nonbeliever is not a scientist but the city dweller who has never had any real contact with nature and little experience of natural rhythms.

Camille Flammarion, the French astronomer, testifies to the success of Moon planting, though:

"Cucumbers increase at Full Moon, as well as radishes, turnips, leeks, lilies, horseradish, and saffron; onions, on the contrary, are much larger and better nourished during the decline and old age of the Moon than at its increase, during its youth and fullness, which is the reason the Egyptians abstained from onions, on account of their antipathy to the Moon. Herbs gathered while the Moon increases are of great efficiency. If the vines are trimmed at night when the Moon is in the sign of the Lion, Sagittarius, the Scorpion, or the Bull, it will save them from field rats, moles, snails, flies, and other animals."

Dr. Clark Timmins is one of the few modern scientists to have conducted tests in Moon planting. Following is a summary of his experiments:

Beets: When sown with the Moon in Scorpio, the germination rate was 71 percent; when sown in Sagittarius, the germination rate was 58 percent.

Scotch marigold: When sown with the Moon in Cancer, the germination rate was 90 percent; when sown in Leo, the rate was 32 percent.

Carrots: When sown with the Moon in Scorpio, the germination rate was 64 percent; when sown in Sagittarius, the germination rate was 47 percent.

Tomatoes: When sown with the Moon in Cancer, the germination rate was 90 percent; but when sown with the Moon in Leo, the germination rate was 58 percent.

Two things should be emphasized. First, remember that this is only a summary of the results of the experiments; the experiments themselves were conducted in a scientific manner to eliminate any variation in soil, temperature, moisture, and so on, so that only the Moon sign is varied. Second, note that these astonishing results were obtained without regard to the phase of the Moon—the other factor we use in Moon planting, and which presumably would have increased the differential in germination rates.

Dr. Timmins also tried transplanting Cancer- and Leo-planted tomato seedlings while the Cancer Moon was waxing. The result was 100 percent survival. When transplanting was done with the waning Sagittarius Moon, there was 0 percent survival. Dr. Timmins' tests show that the Cancer-planted tomatoes had blossoms twelve days earlier than those planted under Leo; the Cancer-planted tomatoes had an average height of twenty inches at that time compared to fifteen inches for the Leo-planted; the first ripe tomatoes were gathered from the Cancer plantings eleven days ahead of the Leo plantings; and a count of the hanging fruit and its size and weight shows an advantage to the Cancer plants over the Leo plants of 45 percent.

Dr. Timmins also observed that there have been similar tests that did not indicate results favorable to the Moon planting theory. As a scientist, he asked why one set of experiments indicated a positive verification of Moon planting, and others did not. He checked these other tests and found that the experimenters had not followed the geocentric system for determining the Moon sign positions, but the heliocentric. When the times used in these other tests were converted to the geocentric system, the dates chosen often were found to be in barren, rather than fertile, signs. Without going into a technical explanation, it is sufficient to point out that geocentric and heliocentric positions often vary by as much as four days. This is a large enough differential to place the Moon in Cancer, for example, in the heliocentric system, and at the same time in Leo by the geocentric system.

Most almanacs and calendars show the Moon's signs heliocentrically—and thus incorrectly for Moon planting—while the *Moon Sign Book* is calculated correctly for planting purposes, using the geocentric system. Some readers are confused because the *Moon Sign Book* talks about first, second, third, and fourth quarters, while other almanacs refer to these same divisions as New Moon, first quarter, Full Moon, and fourth quarter. Thus the almanacs say first quarter when the *Moon Sign Book* says second quarter.

There is nothing complicated about using astrology in agriculture and horticulture in order to increase both pleasure and profit, but there is one very important rule that is often neglected—use common sense! Of course this is one rule that should be remembered in every activity we undertake, but in the case of gardening and farming by the Moon, if it is not possible to use the best dates for planting or harvesting, we must select the next best and just try to do the best we can.

This brings up the matter of the other factors to consider in your gardening work. The dates we give as best for a certain activity apply to the entire country (with slight time correction), but in your section of the country you may be buried under three feet of snow on a date we say is good to plant your flowers. So we have factors of weather, season, temperature, and moisture variations, soil conditions, your own available time and opportunity, and so forth. Some astrologers like to think it is all a matter of science, but gardening is also an art. In art, you develop an instinctive identification with your work and influence it with your feelings and wishes.

The *Moon Sign Book* gives you the place of the Moon for every day of the year so that you can select the best times once you have become familiar with the rules and practices of lunar agriculture. We give you specific, easy-to-follow directions so that you can get right down to work.

We give you the best dates for planting, and also for various related activities, including cultivation, fertilizing, harvesting, irrigation, and getting rid of weeds and pests. But we cannot tell you exactly when it's good to plant. Many of these rules were learned by observation and experience; as the body of experience grew, we could see various patterns emerging that allowed us to make judgments about new things. That's what you should do, too. After you have worked with lunar agriculture for a while and have gained a working knowledge, you will probably begin to try new things—and we hope you will share your experiments and findings with us. That's how the science grows.

Here's an example of what we mean. Years ago Llewellyn George suggested that we try to combine our bits of knowledge about what to expect in planting under each of the Moon signs in order to benefit from several lunar factors in one plant. From this came our rule for developing "thoroughbred seed." To develop thoroughbred seed, save the seed for three successive years from plants grown by the correct Moon sign and phase. You can plant in the first quarter phase and in the sign of Cancer for fruitfulness; the second year, plant seeds from the first year plants in Libra for beauty; and in the third year, plant the seeds from the second year plants in Taurus to produce hardiness. In a similar manner you can combine the fruitfulness of Cancer, the good root growth of Pisces, and the sturdiness and good vine growth of Scorpio. And don't forget the characteristics of Capricorn: hardy like Taurus, but drier and perhaps more resistant to drought and disease.

Unlike common almanacs, we consider both the Moon's phase and the Moon's sign in making our calculations for the proper timing of our work. It is perhaps a little easier to understand this if we remind you that we are all living in the center of a vast electromagnetic field that is the Earth and its environment in space. Everything that occurs within this electromagnetic field has an effect on everything else within the field. The Moon and the Sun are the most important of the factors affecting the life of the Earth, and it is their relative positions to the Earth that we project for each day of the year.

Many people claim that not only do they achieve larger crops gardening by the Moon, but that their fruits and vegetables are much tastier. A number of organic gardeners have also become lunar gardeners using the natural rhythm of life forces that we experience through the relative movements of the Sun and Moon. We provide a few basic rules and then give you day-by-day guidance for your gardening work. You will be able to choose the best dates to meet your own needs and opportunities.

Planting by the Moon's Phases

During the increasing or waxing light—from New Moon to Full Moon—plant annuals that produce their yield above the ground. An annual is a plant that completes its entire life cycle within one growing season and has to be seeded each year. During the decreasing or waning light—from Full Moon to New Moon—plant biennials, perennials, and bulb and root plants. Biennials include crops that are planted one season to winter over and produce crops the next, such as winter wheat. Perennials and bulb and root plants include all plants that grow from the same root each year.

A simpler, less-accurate rule is to plant crops that produce above the ground during the waxing Moon, and to plant crops that produce below the ground during the waning Moon. Thus the old adage, "Plant potatoes during the dark of the Moon." Llewellyn George's system divided the lunar month into quarters. The first two from New Moon to Full Moon are the first and second quarters, and the last two from Full Moon to New Moon the third and fourth quarters. Using these divisions, we can increase our accuracy in timing our efforts to coincide with natural forces.

First Quarter

Plant annuals producing their yield above the ground, which are generally of the leafy kind that produce their seed outside the fruit. Some examples are asparagus, broccoli, brussels sprouts, cabbage, cauliflower, celery, cress, endive, kohlrabi, lettuce, parsley, and spinach. Cucumbers are an exception, as they do best in the first quarter rather than the second, even though the seeds are inside the fruit. Also plant cereals and grains.

Second Quarter

Plant annuals producing their yield above the ground, which are generally of the viney kind that produce their seed inside the fruit. Some examples include beans, eggplant, melons, peas, peppers, pumpkins, squash, tomatoes, etc. These are not hard-and-fast divisions. If you can't plant during the first quarter, plant during the second, and vice versa. There are many plants that seem to do equally well planted in either quarter, such as watermelon, hay, and cereals and grains.

Third Quarter

Plant biennials, perennials, bulbs, root plants, trees, shrubs, berries, grapes, strawberries, beets, carrots, onions, parsnips, rutabagas, potatoes, radishes, peanuts, rhubarb, turnips, winter wheat, etc.

Fourth Quarter

This is the best time to cultivate, turn sod, pull weeds, and destroy pests of all kinds, especially when the Moon is in Aries, Leo, Virgo, Gemini, Aquarius, and Sagittarius.

The Moon in the Signs

Moon in Aries

Barren, dry, fiery, and masculine. Use for destroying noxious weeds.

Moon in Taurus

Productive, moist, earthy, and feminine. Use for planting many crops when hardiness is important, particularly root crops. Also used for lettuce, cabbage, and similar leafy vegetables.

Moon in Gemini

Barren and dry, airy and masculine. Use for destroying noxious growths, weeds, and pests, and for cultivation.

Moon in Cancer

Fruitful, moist, feminine. Use for planting and irrigation.

Moon in Leo

Barren, dry, fiery, masculine. Use for killing weeds or cultivation.

Moon in Virgo

Barren, moist, earthy, and feminine. Use for cultivation and destroying weeds and pests.

Moon in Libra

Semi-fruitful, moist, and airy. Use for planting crops that need good pulp growth. A very good sign for flowers and vines. Also used for seeding hay, corn fodder, and the like.

Moon in Scorpio

Very fruitful and moist, watery and feminine. Nearly as productive as Cancer; use for the same purposes. Especially good for vine growth and sturdiness.

Moon in Sagittarius

Barren and dry, fiery and masculine. Use for planting onions, seeding hay, and for cultivation.

Moon in Capricorn

Productive and dry, earthy and feminine. Use for planting potatoes and other tubers.

Moon in Aquarius

Barren, dry, airy, and masculine. Use for cultivation and destroying noxious growths and pests.

Moon in Pisces

Very fruitful, moist, watery, and feminine. Especially good for root growth.

A Guide to Planting

Plant	Quarter	Sign
Annuals	1st or 2nd	
Apple tree	2nd or 3rd	Cancer, Pisces, Virgo
Artichoke	1st	Cancer, Pisces
Asparagus	1st	Cancer, Scorpio, Pisces
Aster	1st or 2nd	Virgo, Libra
Barley	1st or 2nd	Cancer, Pisces, Libra, Capricorn, Virgo
Beans (bush & pole)	2nd	Cancer, Taurus, Pisces, Libra
Beans (kidney, white, & navy)	1st or 2nd	Cancer, Pisces
Beech tree	2nd or 3rd	Virgo, Taurus
Beets	3rd	Cancer, Capricorn, Pisces, Libra
Biennials	3rd or 4th	
Broccoli	1st	Cancer, Scorpio, Pisces, Libra
Brussels sprouts	1st	Cancer, Scorpio, Pisces, Libra
Buckwheat	1st or 2nd	Capricorn
Bulbs	3rd	Cancer, Scorpio, Pisces
Bulbs for seed	2nd or 3rd	
Cabbage	1st	Cancer, Scorpio, Pisces, Taurus, Libra
Canes (raspberry, blackberry, & gooseberry)	2nd	Cancer, Scorpio, Pisces
Cantaloupe	1st or 2nd	Cancer, Scorpio, Pisces, Taurus, Libra
Carrots	3rd	Cancer, Scorpio, Pisces, Taurus, Libra
Cauliflower	1st	Cancer, Scorpio, Pisces, Libra
Celeriac	3rd	Cancer, Scorpio, Pisces
Celery	1st	Cancer, Scorpio, Pisces
Cereals	1st or 2nd	Cancer, Scorpio, Pisces, Libra
Chard	1st or 2nd	Cancer, Scorpio, Pisces
Chicory	2nd or 3rd	Cancer, Scorpio, Pisces
Chrysanthemum	1st or 2nd	Virgo
Clover	1st or 2nd	Cancer, Scorpio, Pisces

Plant	Quarter	Sign
Coreopsis	2nd or 3rd	Libra
Corn	1st	Cancer, Scorpio, Pisces
Corn for fodder	1st or 2nd	Libra
Cosmo	2nd or 3rd	Libra
Cress	1st	Cancer, Scorpio, Pisces
Crocus	1st or 2nd	Virgo
Cucumber	1st	Cancer, Scorpio, Pisces
Daffodil	1st or 2nd	Libra, Virgo
Dahlia	1st or 2nd	Libra, Virgo
Deciduous trees	2nd or 3rd	Cancer, Scorpio, Pisces, Virgo, Libra
Eggplant	2nd	Cancer, Scorpio, Pisces, Libra
Endive	1st	Cancer, Scorpio, Pisces, Libra
Flowers	1st	Cancer, Scorpio, Pisces, Libra, Taurus, Virgo
Garlic	3rd	Libra, Taurus, Pisces
Gladiola	1st or 2nd	Libra, Virgo
Gourds	1st or 2nd	Cancer, Scorpio, Pisces, Libra
Grapes	2nd or 3rd	Cancer, Scorpio, Pisces, Virgo
Hay	1st or 2nd	Cancer, Scorpio, Pisces, Libra, Taurus
Herbs	1st or 2nd	Cancer, Scorpio, Pisces
Honeysuckle	1st or 2nd	Scorpio, Virgo
Hops	1st or 2nd	Scorpio, Libra
Horseradish	1st or 2nd	Cancer, Scorpio, Pisces
Houseplants	1st	Cancer, Scorpio, Pisces, Libra
Hyacinth	3rd	Cancer, Scorpio, Pisces
Iris	1st or 2nd	Cancer, Virgo
Kohlrabi	1st or 2nd	Cancer, Scorpio, Pisces, Libra
Leek	1st or 2nd	Cancer, Pisces
Lettuce	1st	Cancer, Scorpio, Pisces, Libra, Taurus
Lily	1st or 2nd	Cancer, Scorpio, Pisces
Maple tree	2nd or 3rd	Taurus, Virgo, Cancer, Pisces
Melon	2nd	Cancer, Scorpio, Pisces
Moon vine	1st or 2nd	Virgo

Plant	Quarter	Sign
Morning glory	1st or 2nd	Cancer, Scorpio, Pisces, Virgo
Oak tree	2nd or 3rd	Taurus, Virgo, Cancer, Pisces
Oats	1st or 2nd	Cancer, Scorpio, Pisces, Libra
Okra	1st or 2nd	Cancer, Scorpio, Pisces, Libra
Onion seed	2nd	Cancer, Scorpio, Sagittarius
Onion set	3rd or 4th	Cancer, Pisces, Taurus, Libra
Pansies	1st or 2nd	Cancer, Scorpio, Pisces
Parsley	1st	Cancer, Scorpio, Pisces, Libra
Parsnip	3rd	Cancer, Scorpio, Taurus, Capricorn
Peach tree	2nd or 3rd	Cancer, Taurus, Virgo, Libra
Peanuts	3rd	Cancer, Scorpio, Pisces
Pear tree	2nd or 3rd	Cancer, Scorpio, Pisces, Libra
Peas	2nd	Cancer, Scorpio, Pisces, Libra
Peony	1st or 2nd	Virgo
Peppers	2nd	Cancer, Scorpio, Pisces
Perennials	3rd	
Petunia	1st or 2nd	Libra, Virgo
Plum tree	2nd or 3rd	Cancer, Pisces, Taurus, Virgo
Poppies	1st or 2nd	Virgo
Portulaca	1st or 2nd	Virgo
Potatoes	3rd	Cancer, Scorpio, Libra, Taurus, Capricorn
Privet	1st or 2nd	Taurus, Libra
Pumpkin	2nd	Cancer, Scorpio, Pisces, Libra
Quince	1st or 2nd	Capricorn
Radishes	3rd	Cancer, Scorpio, Pisces, Libra, Capricorn
Rhubarb	3rd	Cancer, Pisces
Rice	1st or 2nd	Scorpio
Roses	1st or 2nd	Cancer, Virgo
Rutabaga	3rd	Cancer, Scorpio, Pisces, Taurus
Saffron	1st or 2nd	Cancer, Scorpio, Pisces
Sage	3rd	Cancer, Scorpio, Pisces

Plant	Quarter	Sign
Salsify	1st	Cancer, Scorpio, Pisces
Shallot	2nd	Scorpio
Spinach	1st	Cancer, Scorpio, Pisces
Squash	2nd	Cancer, Scorpio, Pisces, Libra
Strawberries	3rd	Cancer, Scorpio, Pisces
String beans	1st or 2nd	Taurus
Sunflowers	1st or 2nd	Libra, Cancer
Sweet peas	1st or 2nd	
Tomatoes	2nd	Cancer, Scorpio, Pisces, Capricorn
Trees, shade	3rd	Taurus, Capricorn
Trees, ornamental	2nd	Libra, Taurus
Trumpet vine	1st or 2nd	Cancer, Scorpio, Pisces
Tubers for seed	3rd	Cancer, Scorpio, Pisces, Libra
Tulips	1st or 2nd	Libra, Virgo
Turnips	3rd	Cancer, Scorpio, Pisces, Taurus, Capricorn, Libra
Valerian	1st or 2nd	Virgo, Gemini
Watermelon	1st or 2nd	Cancer, Scorpio, Pisces, Libra
Wheat	1st or 2nd	Cancer, Scorpio, Pisces, Libra

Companion Planting Guide

Plant	Companions	Hindered by
Asparagus	Tomatoes, parsley, basil	None known
Beans	Tomatoes, carrots, cucumbers, garlic, cabbage, beets, corn	Onions, gladiolas
Beets	Onions, cabbage, lettuce, mint, catnip	Pole beans
Broccoli	Beans, celery, potatoes, onions	Tomatoes
Cabbage	Peppermint, sage, thyme, tomatoes	Strawberries, grapes
Carrots	Peas, lettuce, chives, radishes, leeks, onions, sage	Dill, anise
Citrus trees	Guava, live oak, rubber trees, peppers	None known
Corn	Potatoes, beans, peas, melon, squash, pumpkin, sunflowers, soybeans	Quack grass, wheat, straw, mulch
Cucumbers	Beans, cabbage, radishes, sunflowers, lettuce, broccoli, squash	Aromatic herbs
Eggplant	Green beans, lettuce, kale	None known
Grapes	Peas, beans, blackberries	Cabbage, radishes
Melons	Corn, peas	Potatoes, gourds
Onions, leeks	Beets, chamomile, carrots, lettuce	Peas, beans, sage
Parsnip	Peas	None known
Peas	Radishes, carrots, corn, cucumbers, beans, tomatoes, spinach, turnips	Onion, garlic
Potatoes	Beans, corn, peas, cabbage, hemp, cucumbers, eggplant, catnip	Raspberries, pumpkins, tomatoes, sunflowers
Radishes	Peas, lettuce, nasturtiums, cucumbers	Hyssop
Spinach	Strawberries	None known
Squash/Pumpkin	Nasturtiums, corn, mint, catnip	Potatoes
Tomatoes	Asparagus, parsley, chives, onions, carrots, marigolds, nasturtiums, dill	Black walnut roots, fennel, potatoes
Turnips	Peas, beans, brussels sprouts	Potatoes

Plant	Companions	Uses
Anise	Coriander	Flavor candy, pastry, cheeses, cookies
Basil	Tomatoes	Dislikes rue; repels flies and mosquitoes
Borage	Tomatoes, squash	Use in teas
Buttercup	Clover	Hinders delphinium, peonies, monks-hood, columbine

Plant	Companions	Uses
Catnip		Repels flea beetles
Chamomile	Peppermint, wheat, onions, cabbage	Roman chamomile may control damping-off disease; use in herbal sprays
Chervil	Radishes	Good in soups and other dishes
Chives	Carrots	Use in spray to deter black spot on roses
Coriander	Plant anywhere	Hinders seed formation in fennel
Cosmos		Repels corn earworms
Dill	Cabbage	Hinders carrots and tomatoes
Fennel	Plant in borders away from garden	Disliked by all garden plants
Horseradish		Repels potato bugs
Horsetail		Makes fungicide spray
Hyssop		Attracts cabbage fly away from cabbage; harmful to radishes
Lavender	Plant anywhere	Use in spray to control insects on cotton, repels clothes moths
Lovage		Lures horn worms away from tomatoes
Marigolds		Pest repellent; use against Mexican bean beetles and nematodes
Mint	Cabbage, tomatoes	Repels ants, flea beetles, and cabbage worm butterflies
Morning glory	Corn	Helps melon germination
Nasturtiums	Cabbage, cucumbers	Deters aphids, squash bugs, and pumpkin beetles
Okra	Eggplant	Will attract leafhopper (use to trap insects away from other plants)
Parsley	Tomatoes, asparagus	Freeze chopped up leaves to flavor foods
Purslane		Good ground cover
Rosemary		Repels cabbage moths, bean beetles, and carrot flies
Savory		Plant with onions to give them added sweetness
Tansy		Deters Japanese beetles, striped cucumber beetles, and squash bugs
Thyme		Repels cabbage worms
Yarrow		Increases essential oils of neighbors

Moon Void-of-Course

By Kim Rogers-Gallagher

The Moon circles the Earth in about twenty-eight days, moving through each zodiac sign in two-and-a-half days. As she passes through the thirty degrees of each sign, she "visits" with the planets in numerical order, forming aspects with them. Because she moves one degree in just two to two-and-a-half hours, her influence on each planet lasts only a few hours. She eventually reaches the planet that's in the highest degree of any sign and forms what will be her final aspect before leaving the sign. From this point until she enters the next sign, she is referred to as void-of-course.

Think of it this way: the Moon is the emotional "tone" of the day, carrying feelings with her particular to the sign she's "wearing" at the moment. After she has contacted each of the planets, she symbolically "rests" before changing her costume, so her instinct is temporarily on hold. It's during this time that many people feel "fuzzy" or "vague." Plans or decisions made now often do not pan out. Without the instinctual "knowing" the Moon provides as she touches each planet, we tend to be unrealistic or exercise poor judgment. The traditional definition of the void Moon is that "nothing will come of this." Actions initiated under a void Moon are often wasted, irrelevant, or incorrect—usually because information is hidden, missing, or has been overlooked.

Although it's not a good time to initiate plans, routine tasks seem to go along just fine. This period is ideal for reflection. On the lighter side, remember there are good uses for the void Moon. It is the period when the universe seems to be most open to loopholes. It's a great time to make plans you don't want to fulfill or schedule things you don't want to do. See the table on pages 80–85 for a schedule of the Moon's void-of-course times.

Last Aspect   **Moon Enters New Sign**

		January		
1	10:43 am	1	Leo	9:41 pm
3	4:55 pm	3	Virgo	9:52 pm
5	12:25 pm	5	Libra	11:58 pm
8	1:07 am	8	Scorpio	5:00 am
10	10:02 am	10	Sagittarius	1:10 pm
12	9:43 pm	12	Capricorn	11:54 pm
15	4:02 am	15	Aquarius	12:17 pm
17	3:22 pm	18	Pisces	1:17 am
20	1:06 am	20	Aries	1:36 pm
22	2:46 pm	22	Taurus	11:39 pm
24	10:03 pm	25	Gemini	6:11 am
27	1:32 am	27	Cancer	9:01 am
28	11:49 pm	29	Leo	9:10 am
31	1:27 am	31	Virgo	8:23 am
		February		
1	11:17 pm	2	Libra	8:42 am
4	4:27 am	4	Scorpio	11:56 am
6	11:11 am	6	Sagittarius	7:04 pm
8	11:58 pm	9	Capricorn	5:43 am
11	7:39 am	11	Aquarius	6:24 pm
13	11:33 pm	14	Pisces	7:23 am
16	9:32 am	16	Aries	7:30 pm
18	10:52 pm	19	Taurus	5:55 am
21	7:15 am	21	Gemini	1:47 pm
23	12:29 pm	23	Cancer	6:29 pm
25	12:48 pm	25	Leo	8:08 pm
27	3:15 pm	27	Virgo	7:52 pm

Last Aspect   ## Moon Enters New Sign

		March		
1	12:36 pm	1	Libra	7:31 pm
3	3:43 pm	3	Scorpio	9:11 pm
5	11:32 pm	6	Sagittarius	2:36 am
8	6:13 am	8	Capricorn	12:13 pm
10	4:59 pm	11	Aquarius	12:42 am
13	7:57 am	13	Pisces	1:44 pm
15	8:01 pm	16	Aries	2:32 am
18	7:23 am	18	Taurus	12:29 pm
20	3:41 pm	20	Gemini	8:28 pm
22	9:49 pm	23	Cancer	2:16 am
25	12:39 am	25	Leo	5:39 am
27	3:04 am	27	Virgo	6:57 am
29	2:55 am	29	Libra	7:21 am
31	8:13 am	31	Scorpio	8:41 am
		April		
2	8:54 am	2	Sagittarius	12:53 pm
4	4:57 pm	4	Capricorn	9:07 pm
7	4:18 am	7	Aquarius	8:51 am
9	5:44 pm	9	Pisces	9:48 pm
12	8:51 am	12	Aries	9:31 am
14	3:23 pm	14	Taurus	6:55 pm
17	12:57 am	17	Gemini	2:08 am
19	6:21 am	19	Cancer	7:39 am
21	10:07 am	21	Leo	11:42 am
23	11:35 am	23	Virgo	2:24 pm
25	2:21 pm	25	Libra	4:16 pm
27	3:45 pm	27	Scorpio	6:28 pm
29	8:39 pm	29	Sagittarius	10:36 pm

Last Aspect   Moon Enters New Sign

		May		
2	4:08 am	2	Capricorn	6:00 am
4	3:07 pm	4	Aquarius	4:52 pm
7	2:36 am	7	Pisces	5:34 am
9	4:12 pm	9	Aries	5:29 pm
12	12:11 am	12	Taurus	2:48 am
14	8:28 am	14	Gemini	9:18 am
16	1:06 pm	16	Cancer	1:46 pm
18	4:35 pm	18	Leo	5:06 pm
20	7:43 pm	20	Virgo	7:58 pm
22	10:34 pm	22	Libra	10:50 pm
25	12:01 am	25	Scorpio	2:17 am
27	7:13 am	27	Sagittarius	7:15 am
29	12:40 pm	29	Capricorn	2:44 pm
31	11:41 pm	1	Aquarius	1:08 am
		June		
3	10:56 am	3	Pisces	1:34 pm
6	1:49 am	6	Aries	1:50 am
8	9:13 am	8	Taurus	11:41 am
10	3:50 pm	10	Gemini	6:11 pm
12	7:35 pm	12	Cancer	9:50 pm
14	8:38 pm	14	Leo	11:54 pm
16	11:24 pm	17	Virgo	1:41 am
19	1:04 am	19	Libra	4:13 am
21	5:44 am	21	Scorpio	8:14 am
23	11:32 am	23	Sagittarius	2:10 pm
25	7:33 pm	25	Capricorn	10:21 pm
28	5:56 am	28	Aquarius	8:52 am
30	6:03 pm	30	Pisces	9:10 pm

Last Aspect   Moon Enters New Sign

July				
3	7:17 am	3	Aries	9:44 am
5	5:24 pm	5	Taurus	8:29 pm
8	2:10 am	8	Gemini	3:51 am
10	6:17 am	10	Cancer	7:38 am
12	7:48 am	12	Leo	8:53 am
14	6:23 am	14	Virgo	9:15 am
16	9:46 am	16	Libra	10:24 am
18	10:26 am	18	Scorpio	1:42 pm
20	7:43 pm	20	Sagittarius	7:48 pm
23	12:50 am	23	Capricorn	4:39 am
25	10:20 am	25	Aquarius	3:38 pm
27	11:46 pm	28	Pisces	4:00 am
29	11:44 pm	30	Aries	4:42 pm
August				
1	11:54 pm	2	Taurus	4:13 am
4	8:44 am	4	Gemini	12:54 pm
6	5:22 pm	6	Cancer	5:50 pm
7	2:46 pm	8	Leo	7:23 pm
10	3:10 pm	10	Virgo	7:01 pm
11	8:04 pm	12	Libra	6:43 pm
14	4:06 pm	14	Scorpio	8:26 pm
17	1:24 am	17	Sagittarius	1:34 am
19	9:58 am	19	Capricorn	10:17 am
21	9:08 pm	21	Aquarius	9:37 pm
24	4:29 am	24	Pisces	10:11 am
26	10:00 pm	26	Aries	10:49 pm
29	4:47 am	29	Taurus	10:35 am
31	7:13 pm	31	Gemini	8:19 pm

Last Aspect		Moon Enters New Sign			
		September			
3	1:40 am	3	Cancer	2:50 am	
5	4:31 am	5	Leo	5:45 am	
7	4:17 am	7	Virgo	5:53 am	
9	4:59 am	9	Libra	5:01 am	
11	1:16 am	11	Scorpio	5:21 am	
13	7:53 am	13	Sagittarius	8:52 am	
15	2:52 pm	15	Capricorn	4:30 pm	
18	1:13 am	18	Aquarius	3:35 am	
20	9:09 am	20	Pisces	4:15 pm	
23	1:52 am	23	Aries	4:47 am	
25	9:12 am	25	Taurus	4:17 pm	
27	11:03 pm	28	Gemini	2:10 am	
30	6:37 am	30	Cancer	9:46 am	
		October			
2	11:21 am	2	Leo	2:21 pm	
4	9:52 am	4	Virgo	4:00 pm	
6	12:43 pm	6	Libra	3:52 pm	
8	9:38 am	8	Scorpio	3:52 pm	
10	2:27 pm	10	Sagittarius	6:09 pm	
12	8:08 pm	13	Capricorn	12:17 am	
15	5:49 am	15	Aquarius	10:24 am	
17	2:49 pm	17	Pisces	10:52 pm	
20	6:25 am	20	Aries	11:23 am	
22	9:37 pm	22	Taurus	10:30 pm	
25	3:49 am	25	Gemini	7:47 am	
27	10:19 am	27	Cancer	3:14 pm	
29	3:48 pm	29	Leo	8:39 pm	
31	5:01 pm	31	Virgo	11:51 pm	

Last Aspect   Moon Enters New Sign

November				
2	8:36 pm	3	Libra	1:19 am
4	7:34 pm	5	Scorpio	2:16 am
6	11:44 pm	7	Sagittarius	3:28 am
9	7:35 am	9	Capricorn	8:37 am
11	2:57 pm	11	Aquarius	5:32 pm
14	1:33 am	14	Pisces	5:24 am
16	11:37 am	16	Aries	5:59 pm
19	12:33 am	19	Taurus	5:04 am
21	12:27 pm	21	Gemini	1:46 pm
23	4:56 pm	23	Cancer	8:14 pm
25	10:44 pm	26	Leo	1:01 am
28	3:30 am	28	Virgo	4:34 am
30	6:17 am	30	Libra	7:15 am
December				
2	3:08 am	2	Scorpio	9:44 am
4	7:13 am	4	Sagittarius	12:59 pm
6	4:46 pm	6	Capricorn	6:16 pm
8	8:07 pm	9	Aquarius	2:30 am
11	6:09 am	11	Pisces	1:41 pm
13	7:35 pm	14	Aries	2:15 am
16	6:41 am	16	Taurus	1:49 pm
18	4:37 pm	18	Gemini	10:37 pm
21	3:13 am	21	Cancer	4:22 am
23	2:25 am	23	Leo	7:51 am
25	4:28 am	25	Virgo	10:14 am
27	7:21 am	27	Libra	12:38 pm
29	10:05 am	29	Scorpio	3:49 pm
31	2:57 pm	31	Sagittarius	8:21 pm

The Moon's Rhythm

The Moon journeys around Earth in an elliptical orbit that takes about 27.33 days, which is known as a sidereal month (period of revolution of one body about another). She can move up to 15 degrees or as few as 11 degrees in a day, with the fastest motion occurring when the Moon is at perigee (closest approach to Earth). The Moon is never retrograde, but when her motion is slow, the effect is similar to a retrograde period.

Astrologers have observed that people born on a day when the Moon is fast will process information differently from those who are born when the Moon is slow in motion. People born when the Moon is fast process information quickly and tend to react quickly, while those born during a slow Moon will be more deliberate.

The time from New Moon to New Moon is called the synodic month (involving a conjunction), and the average time span between this Sun-Moon alignment is 29.53 days. Since 29.53 won't divide into 365 evenly, we can have a month with two Full Moons or two New Moons.

Moon Aspects

The aspects the Moon will make during the times you are considering are also important. A trine or sextile, and sometimes a conjunction, are considered favorable aspects. A trine or sextile between the Sun and Moon is an excellent foundation for success. Whether or not a conjunction is considered favorable depends upon the planet the Moon is making a conjunction to. If it's joining the Sun, Venus, Mercury, Jupiter, or even Saturn, the aspect is favorable. If the Moon joins Pluto or Mars, however, that would not be considered favorable. There may be exceptions, but it would depend on what you are electing to do. For example, a trine to Pluto might hasten the end of a relationship you want to be free of.

It is important to avoid times when the Moon makes an aspect to or is conjoining any retrograde planet, unless, of course, you want the thing started to end in failure.

After the Moon has completed an aspect to a planet, that planetary energy has passed. For example, if the Moon squares Saturn at 10:00 am, you can disregard Saturn's influence on your activity if it will occur after that time. You should always look ahead at aspects the Moon will make on the day in question, though, because if the Moon opposes Mars at 11:30 pm on that day, you can expect events that stretch into the evening to be affected by the Moon-Mars aspect. A testy conversation might lead to an argument, or more.

Moon Signs

Much agricultural work is ruled by earth signs—Virgo, Capricorn, and Taurus; and the air signs—Gemini, Aquarius, and Libra—rule flying and intellectual pursuits.

Each planet has one or two signs in which its characteristics are enhanced or "dignified," and the planet is said to "rule" that sign. The Sun rules Leo and the Moon rules Cancer, for example. The ruling planet for each sign is listed below. These should not be considered complete lists. We recommend that you purchase a book of planetary rulerships for more complete information.

Aries Moon

The energy of an Aries Moon is masculine, dry, barren, and fiery. Aries provides great start-up energy, but things started at this time may be the result of impulsive action that lacks research or necessary support. Aries lacks staying power.

Use this assertive, outgoing Moon sign to initiate change, but have a plan in place for someone to pick up the reins when you're impatient to move on to the next thing. Work that requires skillful, but not necessarily patient, use of tools—hammering, cutting down trees, etc.—is appropriate in Aries. Expect things to occur rapidly but to also quickly pass. If you are prone to injury or accidents, exercise caution and good judgment in Aries-related activities.

RULER: Mars
IMPULSE: Action
RULES: Head and face

Taurus Moon

A Taurus Moon's energy is feminine, semi-fruitful, and earthy. The Moon is exalted—very strong—in Taurus. Taurus is known as the farmer's sign because of its associations with farmland and precipitation that is the typical day-long "soaker" variety. Taurus energy is good to incorporate into your plans when patience, practicality, and perseverance are needed. Be aware, though, that you may also experience stubbornness in this sign.

Things started in Taurus tend to be long lasting and to increase in value. This can be very supportive energy in a marriage election. On the downside, the fixed energy of this sign resists change or the letting go of even the most difficult situations. A divorce following a marriage that occurred during a Taurus Moon may be difficult and costly to end. Things begun now tend to become habitual and hard to alter. If you want to make changes in something you started, it would be better to wait for Gemini. This is a good time to get a loan, but expect the people in charge of money to be cautious and slow to make decisions.

RULER: Venus

IMPULSE: Stability

RULES: Neck, throat, and voice

Gemini Moon

A Gemini Moon's energy is masculine, dry, barren, and airy. People are more changeable than usual and may prefer to follow intellectual pursuits and play mental games rather than apply themselves to practical concerns.

This sign is not favored for agricultural matters, but it is an excellent time to prepare for activities, to run errands, and write letters. Plan to use a Gemini Moon to exchange ideas, meet people, go on vacations that include walking or biking, or be in situations that require versatility and quick thinking on your feet.

RULER: Mercury

IMPULSE: Versatility

RULES: Shoulders, hands, arms, lungs, and nervous system

Cancer Moon

A Cancer Moon's energy is feminine, fruitful, moist, and very strong. Use this sign when you want to grow things—flowers, fruits, vegetables, commodities, stocks, or collections—for example. This sensitive sign stimulates rapport between people. Considered the most fertile of the signs, it is often associated with mothering. You can use this moontime to build personal friendships that support mutual growth.

Cancer is associated with emotions and feelings. Prominent Cancer energy promotes growth, but it can also turn people pouty and prone to withdrawing into their shells.

RULER: The Moon

IMPULSE: Tenacity

RULES: Chest area, breasts, and stomach

Leo Moon

A Leo Moon's energy is masculine, hot, dry, fiery, and barren. Use it whenever you need to put on a show, make a presentation, or entertain colleagues or guests. This is a proud yet playful energy that exudes self-confidence and is often associated with romance.

This is an excellent time for fund-raisers and ceremonies or to be straightforward, frank, and honest about something. It is advisable not to put yourself in a position of needing public approval or where you might have to cope with underhandedness, as trouble in these areas can bring out the worst Leo traits. There is a tendency in this sign to become arrogant or self-centered.

RULER: The Sun

IMPULSE: I am

RULES: Heart and upper back

Virgo Moon

A Virgo Moon is feminine, dry, barren, earthy energy. It is favorable for anything that needs painstaking attention—especially those things where exactness rather than innovation is preferred.

Use this sign for activities when you must analyze information or when you must determine the value of something. Virgo is the sign of bargain hunting. It's friendly toward agricultural matters with an emphasis on animals and harvesting vegetables. It is an excellent time to care for animals, especially training them and veterinary work.

This sign is most beneficial when decisions have already been made and now need to be carried out. The inclination here is to see details rather than the bigger picture.

There is a tendency in this sign to overdo. Precautions should be taken to avoid becoming too dull from all work and no play. Build a little relaxation and pleasure into your routine from the beginning.

RULER: Mercury

IMPULSE: Discriminating

RULES: Abdomen and intestines

Libra Moon

A Libra Moon's energy is masculine, semi-fruitful, and airy. This energy will benefit any attempt to bring beauty to a place or thing. Libra is considered good energy for starting things of an intellectual nature. Libra is the sign of partnership and unions, which make it an excellent time to form partnerships of any kind, to make agreements, and to negotiate. Even though this sign is good for initiating things, it is crucial to work with a partner who will provide incentive and encouragement, however. A Libra Moon accentuates teamwork (particularly teams of two) and artistic work (especially work that involves color). Make use of this sign when you are decorating your home or shopping for better quality clothing.

RULER: Venus

IMPULSE: Balance

RULES: Lower back, kidneys, and buttocks

Scorpio Moon

The Scorpio Moon is feminine, fruitful, cold, and moist. It is useful when intensity (that sometimes borders on obsession) is needed. Scorpio is considered a very psychic sign. Use this Moon sign when you must back up something you strongly believe in, such as union or employer relations. There is strong group loyalty here, but a Scorpio Moon is also a good time to end connections thoroughly. This is also a good time to conduct research.

The desire nature is so strong here that there is a tendency to manipulate situations to get what one wants or to not see one's responsibility in an act.

RULER: Pluto, Mars (traditional)

IMPULSE: Transformation

RULES: Reproductive organs, genitals, groin, and pelvis

Sagittarius Moon

The Moon's energy is masculine, dry, barren, and fiery in Sagittarius, encouraging flights of imagination and confidence in the flow of life. Sagittarius is the most philosophical sign. Candor and honesty are enhanced when the Moon is here. This is an excellent time to "get things off your chest" and to deal with institutions of higher learning, publishing companies, and the law. It's also a good time for sport and adventure.

Sagittarians are the crusaders of this world. This is a good time to tackle things that need improvement, but don't try to be the diplomat while influenced by this energy. Opinions can run strong, and the tendency to proselytize is increased.

RULER: Jupiter

IMPULSE: Expansion

RULES: Thighs and hips

Capricorn Moon

In Capricorn the Moon's energy is feminine, semi-fruitful, and earthy. Because Cancer and Capricorn are polar opposites, the Moon's energy is thought to be weakened here. This energy encourages the need for structure, discipline, and organization. This is a good time to set goals and plan for the future, tend to family business, and to take care of details requiring patience or a businesslike manner. Institutional activities are favored. This sign should be avoided if you're seeking favors, as those in authority can be insensitive under this influence.

RULER: Saturn

IMPULSE: Ambitious

RULES: Bones, skin, and knees

Aquarius Moon

An Aquarius Moon's energy is masculine, barren, dry, and airy. Activities that are unique, individualistic, concerned with humanitarian issues, society as a whole, and making improvements are favored under this Moon. It is this quality of making improvements that has caused this sign to be associated with inventors and new inventions.

An Aquarius Moon promotes the gathering of social groups for friendly exchanges. People tend to react and speak from an intellectual rather than emotional viewpoint when the Moon is in this sign.

RULER: Uranus and Saturn

IMPULSE: Reformer

RULES: Calves and ankles

Pisces Moon

A Pisces Moon is feminine, fruitful, cool, and moist. This is an excellent time to retreat, meditate, sleep, pray, or make that dreamed-of escape into a fantasy vacation. However, things are not always what they seem to be with the Moon in Pisces. Personal boundaries tend to be fuzzy, and you may not be seeing things clearly. People tend to be idealistic under this sign, which can prevent them from seeing reality.

There is a live and let live philosophy attached to this sign, which in the idealistic world may work well enough, but chaos is frequently the result. That's why this sign is also associated with alcohol and drug abuse, drug trafficking, and counterfeiting. On the lighter side, many musicians and artists are ruled by Pisces. It's only when they move too far away from reality that the dark side of substance abuse, suicide, or crime takes away life.

RULER: Jupiter and Neptune

IMPULSE: Empathetic

RULES: Feet

More About Zodiac Signs

Element (Triplicity)

Each of the zodiac signs is classified as belonging to an element; these are the four basic elements:

Fire Signs

Aries, Sagittarius, and Leo are action-oriented, outgoing, energetic, and spontaneous.

Earth Signs

Taurus, Capricorn, and Virgo are stable, conservative, practical, and oriented to the physical and material realm.

Air Signs

Gemini, Aquarius, and Libra are sociable and critical, and they tend to represent intellectual responses rather than feelings.

Water Signs

Cancer, Scorpio, and Pisces are emotional, receptive, intuitive, and can be very sensitive.

Quality (Quadruplicity)

Each zodiac sign is further classified as being cardinal, mutable, or fixed. There are four signs in each quadruplicity, one sign from each element.

Cardinal Signs

Aries, Cancer, Libra, and Capricorn represent beginnings and newly initiated action. They initiate each new season in the cycle of the year.

Fixed Signs

Taurus, Leo, Scorpio, and Aquarius want to maintain the status quo through stubbornness and persistence; they represent that "between" time. For example, Leo is the month when summer really feels like summer.

Mutable Signs

Pisces, Gemini, Virgo, and Sagittarius adapt to change and tolerate situations. They represent the last month of each season, when things are changing in preparation for the coming season.

Nature and Fertility

In addition to a sign's element and quality, each sign is further classified as either fruitful, semi-fruitful, or barren. This classification is the most important for readers who use the gardening information in the *Moon Sign Book* because the timing of most events depends on the fertility of the sign occupied by the Moon. The water signs of Cancer, Scorpio, and Pisces are the most fruitful. The semi-fruitful signs are the earth signs Taurus and Capricorn, and the air sign Libra. The barren signs correspond to fire-signs Aries, Leo, and Sagittarius; air-signs Gemini and Aquarius; and earth-sign Virgo.

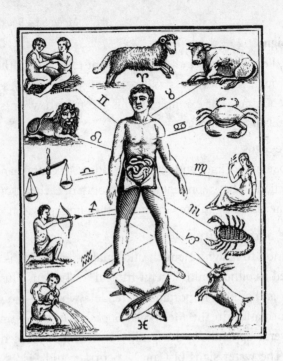

Good Timing

By *Sharon Leah*

Electional astrology is the art of electing times to begin any undertaking. Say, for example, you want to start a business. That business will experience ups and downs, as well as reach its potential, according to the promise held in the universe at the time the business was started—its birth time. The horoscope (birth chart) set for the date, time, and place that a business starts would indicate the outcome—its potential to succeed.

So, you might ask yourself the question: If the horoscope for a business start can show success or failure, why not begin at a time that is more favorable to the venture? Well, you can.

While no time is perfect, there are better times and better days to undertake specific activities. There are thousands of examples that prove electional astrology is not only practical, but that it can make a difference in our lives. There are rules for electing times to begin various activities—even shopping. You'll find detailed instructions about how to make elections beginning on page 111.

Personalizing Elections

The election rules in this almanac are based upon the planetary positions at the time for which the election is made. They do not depend on any type of birth chart. However, a birth chart based upon the time, date, and birthplace of an event has advantages. No election is effective for every person. For example, you may leave home to begin a trip at the same time as a friend, but each of you will have a different experience according to whether or not your birth chart favors the trip.

Not all elections require a birth chart, but the timing of very important events—business starts, marriages, etc.—would benefit from the additional accuracy a birth chart provides. To order a birth chart for yourself or a planned event, visit our Web site at www.llewellyn.com.

Some Things to Consider

You've probably experienced good timing in your life. Maybe you were at the right place at the right time to meet a friend whom you hadn't seen in years. Frequently, when something like that happens, it is the result of following an intuitive impulse—that "gut instinct." Consider for a moment that you were actually responding to planetary energies. Electional astrology is a tool that can help you to align with energies, present and future, that are available to us through planetary placements.

Significators

Decide upon the important significators (planet, sign, and house ruling the matter) for which the election is being made. The Moon is the most important significator in any election, so the

Moon should always be fortified (strong by sign and making favorable aspects to other planets). The Moon's aspects to other planets are more important than the sign the Moon is in.

Other important considerations are the significators of the Ascendant and Midheaven—the house ruling the election matter and the ruler of the sign on that house cusp. Finally, any planet or sign that has a general rulership over the matter in question should be taken into consideration.

Nature and Fertility

Determine the general nature of the sign that is appropriate for your election. For example, much agricultural work is ruled by the earth signs of Virgo, Capricorn, and Taurus; while the air signs—Gemini, Aquarius, and Libra—rule intellectual pursuits.

One Final Comment

Use common sense. If you must do something, like plant your garden or take an airplane trip on a day that doesn't have the best aspects, proceed anyway, but try to minimize problems. For example, leave early for the airport to avoid being left behind due to delays in the security lanes. When you have no other choice, do the best that you can under the circumstances at the time.

If you want to personalize your elections, please turn to page 111 for more information. If you want a quick and easy answer, you can refer to Llewellyn's Astro Almanac.

Llewellyn's Astro Almanac

The Astro Almanac tables, beginning on the next page, can help you find the dates best suited to particular activities. The dates provided are determined from the Moon's sign, phase, and aspects to other planets. Please note that the Astro Almanac does not take personal factors, such as your Sun and Moon sign, into account. The dates are general, and they will apply for everyone. Some activities will not have suitable dates during a particular month, so no dates will be shown.

Activity	January
Animals (Neuter or spay)	10, 11, 12, 13, 14, 15, 18, 19, 20
Animals (Sell or buy)	25
Automobile (Buy)	4, 13, 14, 25
Brewing	1, 9, 10
Build (foundation)	15, 18
Business (Conducting for Self and Others)	4, 9, 20, 25
Business (Start new)	23
Can Fruits and Vegetables	1, 9
Can Preserves	1, 9
Concrete (Pour)	2, 3, 30
Construction (Begin new)	4, 8, 12, 20, 23, 25, 27
Consultants (Begin work with)	4, 8, 12, 13, 23, 27
Contracts (Bid on)	4, 12, 13, 18, 27
Cultivate	
Decorating	15, 16, 17, 25, 26, 27
Demolition	2, 10, 11, 30
Electronics (Buy)	
Entertain Guests	23, 24, 25
Floor Covering (Laying new)	2, 3, 4, 5, 6, 7, 8, 30, 31
Habits (Break)	
Hair (Cut)	13, 14, 18, 19, 22, 24, 25, 26
Harvest (Grain for storage)	3, 31
Harvest (Root crops)	2, 3, 10, 11, 12
Investments (New)	4, 25
Loan (Ask for)	23, 24
Massage (Relaxing)	
Mow Lawn (Decrease growth)	2, 3, 4, 5, 6, 7, 10, 11, 12, 13, 31
Mow Lawn (Increase growth)	16, 17, 18, 19, 22, 23, 24, 25, 26, 27, 28
Mushrooms (Pick)	29, 30, 31
Negotiate (Business for the Elderly)	10, 25, 29
Prune for Better Fruit	8, 9, 10, 11, 12
Prune to Promote Healing	13, 14
Wean Children	10, 11, 12, 13, 14
Wood Floors (Installing)	13, 14
Write Letters or Contracts	1, 10, 13, 14, 24, 28

Activity	February
Animals (Neuter or spay)	7, 8, 9, 11, 14, 15, 16
Animals (Sell or buy)	20, 24
Automobile (Buy)	1, 11, 21, 23
Brewing	5, 6
Build (foundation)	21
Business (Conducting for Self and Others)	3, 8, 19, 24
Business (Start new)	19, 28
Can Fruits and Vegetables	5, 6
Can Preserves	5, 6
Concrete (Pour)	12
Construction (Begin new)	3, 8, 9, 19, 24
Consultants (Begin work with)	1, 9, 11, 17, 29, 23, 24
Contracts (Bid on)	1, 9, 14, 18, 23, 24
Cultivate	12, 13
Decorating	13, 21, 22, 23
Demolition	7, 8
Electronics (Buy)	11, 23
Entertain Guests	3, 19, 20
Floor Covering (Laying new)	1, 2, 3, 4, 11, 12, 13
Habits (Break)	9, 11, 12
Hair (Cut)	14, 15, 21, 25
Harvest (Grain for storage)	
Harvest (Root crops)	7, 8, 11, 12
Investments (New)	3, 24
Loan (Ask for)	19, 20, 21, 26, 27
Massage (Relaxing)	3, 20, 24
Mow Lawn (Decrease growth)	1, 2, 3, 7, 8, 9, 11, 12
Mow Lawn (Increase growth)	19, 20, 21, 22, 23, 24, 25
Mushrooms (Pick)	27, 28
Negotiate (Business for the Elderly)	21
Prune for Better Fruit	4, 5, 6, 7, 8
Prune to Promote Healing	9, 11
Wean Children	7, 8, 9, 11, 12
Wood Floors (Installing)	9, 11
Write Letters or Contracts	6, 11, 21, 25

Activity	March
Animals (Neuter or spay)	7, 8, 9, 10, 13, 14
Animals (Sell or buy)	19, 22, 24, 26
Automobile (Buy)	9, 10, 20, 21
Brewing	4, 14
Build (foundation)	19, 20
Business (Conducting for Self and Others)	4, 10, 20, 25
Business (Start new)	19, 28
Can Fruits and Vegetables	4, 14
Can Preserves	4
Concrete (Pour)	11, 12
Construction (Begin new)	9, 10, 19, 20, 24, 25
Consultants (Begin work with)	9, 19, 21, 24, 26
Contracts (Bid on)	7, 9, 14, 21, 24, 26
Cultivate	6, 7, 11, 12
Decorating	20, 21, 22, 29
Demolition	6, 7
Electronics (Buy)	21
Entertain Guests	19, 20, 26
Floor Covering (Laying new)	1, 2, 3, 11, 12, 30, 31
Habits (Break)	9, 10, 11, 12, 13
Hair (Cut)	9, 10, 13, 14, 15, 19, 20, 21, 22
Harvest (Grain for storage)	6, 7
Harvest (Root crops)	6, 7, 8, 11, 12, 13
Investments (New)	4, 25
Loan (Ask for)	19, 20, 25
Massage (Relaxing)	26
Mow Lawn (Decrease growth)	1, 2, 6, 7, 8, 9, 10, 11, 12, 30
Mow Lawn (Increase growth)	18, 19, 20, 21, 22, 23, 24
Mushrooms (Pick)	1, 28, 29, 30
Negotiate (Business for the Elderly)	7, 20, 22
Prune for Better Fruit	3, 4, 5, 6, 31
Prune to Promote Healing	8, 9, 10
Wean Children	6, 7, 8, 9, 10, 11, 12, 13
Wood Floors (Installing)	8, 9, 10
Write Letters or Contracts	5, 10, 15, 20, 25

Activity	April
Animals (Neuter or spay)	6, 7, 10, 11, 12
Animals (Sell or buy)	16, 20, 21
Automobile (Buy)	5, 7, 24, 25
Brewing	1, 2, 10, 11, 29
Build (foundation)	16, 17
Business (Conducting for Self and Others)	3, 8, 19, 23
Business (Start new)	16, 17, 25
Can Fruits and Vegetables	1, 10, 11, 29
Can Preserves	1, 29
Concrete (Pour)	8, 9
Construction (Begin new)	3, 6, 8, 16, 19, 20, 23
Consultants (Begin work with)	5, 6, 15, 16, 20, 24
Contracts (Bid on)	5, 6, 10, 11, 20, 24
Cultivate	2, 3, 4, 7, 8, 9, 12, 13, 29, 30
Decorating	17, 18, 19, 25, 26, 27
Demolition	2, 3, 12, 29, 30
Electronics (Buy)	
Entertain Guests	15, 16, 17, 25
Floor Covering (Laying new)	7, 8, 9, 12
Habits (Break)	7, 8, 9, 12
Hair (Cut)	2, 5, 6, 10, 11, 14, 16, 17, 18, 21, 29
Harvest (Grain for storage)	3, 4, 30
Harvest (Root crops)	3, 4, 7, 8, 9, 12, 30
Investments (New)	3, 23
Loan (Ask for)	15, 16, 21, 22, 23
Massage (Relaxing)	16, 25
Mow Lawn (Decrease growth)	2, 3, 4, 5, 6, 7, 8, 9, 12, 29, 30
Mow Lawn (Increase growth)	15, 16, 17, 19, 20, 21, 25, 26
Mushrooms (Pick)	27, 28, 29
Negotiate (Business for the Elderly)	2, 7, 21, 29
Prune for Better Fruit	1, 2, 3, 4, 28, 29, 30
Prune to Promote Healing	5, 6, 7
Wean Children	2, 3, 4, 5, 6, 7, 8, 9, 30
Wood Floors (Installing)	4, 5, 6, 7
Write Letters or Contracts	2, 7, 15, 16, 21, 29

Activity	May
Animals (Neuter or spay)	7, 8, 9
Animals (Sell or buy)	14, 18, 20, 27
Automobile (Buy)	2, 4, 21, 22, 30, 31
Brewing	8, 9
Build (foundation)	19, 20
Business (Conducting for Self and Others)	3, 8, 18, 23
Business (Start new)	14, 22
Can Fruits and Vegetables	8, 9
Can Preserves	12, 13
Concrete (Pour)	5, 6, 12
Construction (Begin new)	3, 4, 14, 18, 23, 31
Consultants (Begin work with)	2, 4, 12, 14, 16, 18, 21, 30, 31
Contracts (Bid on)	2, 4, 7, 9, 16, 18, 21, 30, 31
Cultivate	6, 7, 9, 10, 11, 27, 28, 29
Decorating	14, 15, 16, 23, 24, 25
Demolition	1, 9, 10, 11, 12, 27, 28
Electronics (Buy)	
Entertain Guests	20
Floor Covering (Laying new)	5, 6, 12, 13
Habits (Break)	6, 7, 10, 11
Hair (Cut)	2, 3, 7, 12, 14, 15, 18, 27, 31
Harvest (Grain for storage)	1, 2, 5, 28, 29
Harvest (Root crops)	1, 5, 6, 10, 11, 28, 29
Investments (New)	3, 23
Loan (Ask for)	19, 20
Massage (Relaxing)	5
Mow Lawn (Decrease growth)	1, 2, 3, 4, 5, 6, 9, 10, 11, 12, 28, 29, 30, 31
Mow Lawn (Increase growth)	14, 15, 16, 17, 22, 23, 24
Mushrooms (Pick)	26, 27, 28
Negotiate (Business for the Elderly)	18, 31
Prune for Better Fruit	1, 27
Prune to Promote Healing	2, 3, 4, 30, 31
Wean Children	1, 2, 3, 4, 5, 6, 7, 28, 29, 30, 31
Wood Floors (Installing)	2, 3, 4, 29, 30, 31
Write Letters or Contracts	4, 12, 14, 18, 27

Activity	June
Animals (Neuter or spay)	4
Animals (Sell or buy)	15, 19, 23, 24
Automobile (Buy)	10, 19, 27, 28
Brewing	4
Build (foundation)	15, 16
Business (Conducting for Self and Others)	2, 7, 16, 21
Business (Start new)	26
Can Fruits and Vegetables	4
Can Preserves	9, 10
Concrete (Pour)	1, 2, 9, 10, 29, 30
Construction (Begin new)	2, 7, 10, 15, 16, 23, 28
Consultants (Begin work with)	10, 15, 20, 23, 28
Contracts (Bid on)	5, 6, 10, 15, 20, 23
Cultivate	6, 7, 11
Decorating	12, 19, 20, 21
Demolition	6, 7
Electronics (Buy)	1, 10, 20, 28
Entertain Guests	14, 15, 19, 20
Floor Covering (Laying new)	1, 2, 8, 9, 10, 11, 28, 29
Habits (Break)	6, 7, 8, 11
Hair (Cut)	8, 9, 10, 14, 30
Harvest (Grain for storage)	1, 2, 28, 29, 30
Harvest (Root crops)	1, 2, 6, 7, 11, 28, 29, 30
Investments (New)	2, 21
Loan (Ask for)	14, 15, 16
Massage (Relaxing)	15, 19
Mow Lawn (Decrease growth)	1, 2, 6, 7, 8, 9, 10, 11, 27, 28, 29
Mow Lawn (Increase growth)	13, 14, 19, 20, 25
Mushrooms (Pick)	25, 26, 27
Negotiate (Business for the Elderly)	10, 14, 23
Prune for Better Fruit	
Prune to Promote Healing	26, 27, 28
Wean Children	1, 2, 27, 28, 29, 30
Wood Floors (Installing)	26, 27, 28
Write Letters or Contracts	1, 10, 15, 23, 28

Activity	July
Animals (Neuter or spay)	
Animals (Sell or buy)	12, 21, 24
Automobile (Buy)	8, 9, 14, 16
Brewing	1, 2, 29, 30
Build (foundation)	12, 13
Business (Conducting for Self and Others)	1, 7, 15, 20, 31
Business (Start new)	23
Can Fruits and Vegetables	1, 2, 11, 29, 30
Can Preserves	6, 7, 11
Concrete (Pour)	6, 7, 26, 27
Construction (Begin new)	7, 8, 12, 15, 21, 25, 31
Consultants (Begin work with)	3, 7, 8, 12, 17, 21, 22, 25, 30
Contracts (Bid on)	2, 3, 8, 12, 17, 21, 22, 30
Cultivate	3, 4, 5, 8, 9
Decorating	16, 17, 18, 25
Demolition	3, 4, 30, 31
Electronics (Buy)	8, 17, 25
Entertain Guests	13, 14, 17
Floor Covering (Laying new)	5, 6, 7, 8, 9, 26, 27
Habits (Break)	5, 8
Hair (Cut)	1, 2, 5, 6, 8, 12, 20, 23, 24, 28
Harvest (Grain for storage)	4, 26, 27, 31
Harvest (Root crops)	3, 4, 5, 8, 9, 26, 27, 31
Investments (New)	1, 20, 31
Loan (Ask for)	13, 14
Massage (Relaxing)	
Mow Lawn (Decrease growth)	3, 4, 8, 9, 30, 31
Mow Lawn (Increase growth)	16, 17, 23, 24
Mushrooms (Pick)	24, 25, 26
Negotiate (Business for the Elderly)	20, 22, 25
Prune for Better Fruit	
Prune to Promote Healing	
Wean Children	21, 26, 27
Wood Floors (Installing)	
Write Letters or Contracts	8, 12, 20, 25

Activity	August
Animals (Neuter or spay)	
Animals (Sell or buy)	17, 22, 23
Automobile (Buy)	4, 5, 11
Brewing	7, 8, 25, 26
Build (foundation)	
Business (Conducting for Self and Others)	5, 14, 19, 29
Business (Start new)	19
Can Fruits and Vegetables	7, 8, 25, 26
Can Preserves	3, 7, 8, 30, 31
Concrete (Pour)	3, 30, 31
Construction (Begin new)	4, 5, 9, 14, 17, 19, 22, 29, 31
Consultants (Begin work with)	2, 4, 7, 9, 11, 17, 20, 22, 27, 30, 31
Contracts (Bid on)	4, 7, 11, 17, 20, 27, 31
Cultivate	4, 5, 6, 8, 9
Decorating	12, 13, 14, 22, 23, 24
Demolition	1, 8, 9, 27, 28
Electronics (Buy)	4
Entertain Guests	8, 13, 23
Floor Covering (Laying new)	2, 3, 4, 5, 8, 9, 29, 30, 31
Habits (Break)	6
Hair (Cut)	2, 4, 8, 17, 24, 29, 30, 31
Harvest (Grain for storage)	1, 27, 28
Harvest (Root crops)	1, 5, 6, 27, 28
Investments (New)	19, 29
Loan (Ask for)	10
Massage (Relaxing)	4, 8, 13
Mow Lawn (Decrease growth)	1, 4, 5, 8, 27, 28, 30, 31
Mow Lawn (Increase growth)	12, 13, 14, 19, 20, 21
Mushrooms (Pick)	23, 24, 25
Negotiate (Business for the Elderly)	4, 8, 22
Prune for Better Fruit	
Prune to Promote Healing	
Wean Children	
Wood Floors (Installing)	
Write Letters or Contracts	4, 8, 11, 17, 21, 31

Activity	September
Animals (Neuter or spay)	13
Animals (Sell or buy)	13, 21
Automobile (Buy)	1, 2, 7, 8, 16, 17, 18, 28
Brewing	3, 4
Build (foundation)	12
Business (Conducting for Self and Others)	3, 12, 17, 28
Business (Start new)	9, 18
Can Fruits and Vegetables	3, 4
Can Preserves	3, 4, 26
Concrete (Pour)	6, 26
Construction (Begin new)	3, 5, 17, 18, 27, 28
Consultants (Begin work with)	3, 5, 7, 16, 18, 27
Contracts (Bid on)	3, 5, 7, 16, 18, 23
Cultivate	1, 2, 5, 6, 7
Decorating	9, 10, 11, 18, 19, 20
Demolition	5, 6, 23, 24
Electronics (Buy)	
Entertain Guests	7, 10
Floor Covering (Laying new)	1, 2, 5, 6, 7, 24, 25, 26, 27, 28, 29
Habits (Break)	2, 5, 6, 7
Hair (Cut)	2, 3, 5, 15, 17, 20, 21, 22, 27, 28
Harvest (Grain for storage)	1, 24, 25, 28, 29
Harvest (Root crops)	1, 2, 5, 6, 24, 25, 28, 29
Investments (New)	17, 28
Loan (Ask for)	
Massage (Relaxing)	7
Mow Lawn (Decrease growth)	1, 5, 6, 7, 24, 27, 28, 29
Mow Lawn (Increase growth)	9, 10, 15, 16, 17, 18, 19
Mushrooms (Pick)	22, 23, 24
Negotiate (Business for the Elderly)	5, 13, 18, 28
Prune for Better Fruit	
Prune to Promote Healing	
Wean Children	
Wood Floors (Installing)	
Write Letters or Contracts	5, 7, 13, 18, 27

Activity	October
Animals (Neuter or spay)	11, 12, 13, 14
Animals (Sell or buy)	10, 15, 18
Automobile (Buy)	15, 26
Brewing	1, 2, 28, 29
Build (foundation)	
Business (Conducting for Self and Others)	3, 12, 17, 27
Business (Start new)	15
Can Fruits and Vegetables	1, 28, 29
Can Preserves	1, 23, 24, 28, 29
Concrete (Pour)	3, 4, 23, 24, 30, 31
Construction (Begin new)	2, 3, 12, 15, 17, 24, 27, 29
Consultants (Begin work with)	2, 7, 11, 15, 17, 24, 28, 29
Contracts (Bid on)	2, 11, 15, 20, 28, 29
Cultivate	3, 4, 5, 6, 30, 31
Decorating	7, 8, 16, 17
Demolition	2, 3, 29, 30
Electronics (Buy)	7, 17
Entertain Guests	9, 14, 19
Floor Covering (Laying new)	2, 3, 4, 5, 6, 7, 23, 24, 25, 26, 29, 30, 31
Habits (Break)	4, 31
Hair (Cut)	2, 11, 12, 18, 19, 22, 24, 25, 26, 29
Harvest (Grain for storage)	25, 26, 27, 30
Harvest (Root crops)	3, 4, 25, 26, 27, 30, 31
Investments (New)	17, 27
Loan (Ask for)	
Massage (Relaxing)	1, 28
Mow Lawn (Decrease growth)	2, 3, 4, 5, 6, 25, 26, 29, 30, 31
Mow Lawn (Increase growth)	9, 12, 13, 14, 15, 16
Mushrooms (Pick)	21, 22, 23
Negotiate (Business for the Elderly)	11, 16, 26, 30
Prune for Better Fruit	
Prune to Promote Healing	
Wean Children	29, 30, 31
Wood Floors (Installing)	
Write Letters or Contracts	2, 7, 10, 15, 25, 29

Activity	November
Animals (Neuter or spay)	8, 9, 10, 11
Animals (Sell or buy)	9, 11, 14, 21
Automobile (Buy)	1, 2, 11, 22
Brewing	24
Build (foundation)	12
Business (Conducting for Self and Others)	1, 10, 16, 26, 30
Business (Start new)	11, 21
Can Fruits and Vegetables	5, 24
Can Preserves	5, 24
Concrete (Pour)	26, 27
Construction (Begin new)	1, 10, 11, 21, 25, 26, 30
Consultants (Begin work with)	2, 11, 12, 17, 21, 25, 27
Contracts (Bid on)	2, 11, 16, 17, 25, 27
Cultivate	1, 2, 26, 27, 28, 29, 30
Decorating	12, 13, 14
Demolition	26, 27
Electronics (Buy)	12
Entertain Guests	28
Floor Covering (Laying new)	1, 2, 3, 4, 22, 26, 27, 28, 29, 30
Habits (Break)	
Hair (Cut)	8, 9, 10, 14, 19, 21, 22
Harvest (Grain for storage)	22, 23, 26, 27
Harvest (Root crops)	22, 23, 26, 27
Investments (New)	16, 26
Loan (Ask for)	19, 20
Massage (Relaxing)	14, 28
Mow Lawn (Decrease growth)	1, 2, 3, 4, 22, 25, 26, 27, 28, 29, 30
Mow Lawn (Increase growth)	9, 10, 11, 19, 20
Mushrooms (Pick)	20, 21, 22
Negotiate (Business for the Elderly)	12, 22
Prune for Better Fruit	5
Prune to Promote Healing	
Wean Children	26, 27, 28
Wood Floors (Installing)	
Write Letters or Contracts	6, 11, 21, 25

Activity	December
Animals (Neuter or spay)	5, 6, 7, 8, 11, 12
Animals (Sell or buy)	8, 12, 18
Automobile (Buy)	7, 8, 19, 20, 29
Brewing	3, 4, 22, 30, 31
Build (foundation)	9, 10
Business (Conducting for Self and Others)	10, 16, 25, 30
Business (Start new)	8, 18
Can Fruits and Vegetables	3, 22, 30, 31
Can Preserves	3, 22, 30, 31
Concrete (Pour)	24
Construction (Begin new)	8, 10, 16, 18, 23, 25
Consultants (Begin work with)	7, 8, 16, 18, 23, 24, 28
Contracts (Bid on)	7, 8, 12, 13, 23, 24, 258
Cultivate	23, 24
Decorating	9, 10, 11, 19, 20, 21
Demolition	23, 24
Electronics (Buy)	28
Entertain Guests	10, 16
Floor Covering (Laying new)	1, 2, 23, 24, 25, 26, 27, 28, 29
Habits (Break)	
Hair (Cut)	8, 11, 12, 16, 18, 19, 23
Harvest (Grain for storage)	24, 25
Harvest (Root crops)	23, 24
Investments (New)	16, 25
Loan (Ask for)	16, 17, 18
Massage (Relaxing)	22
Mow Lawn (Decrease growth)	1, 23, 24, 25, 26, 27, 28, 29
Mow Lawn (Increase growth)	7, 8, 9, 16, 17, 18, 19, 20
Mushrooms (Pick)	20, 21, 22
Negotiate (Business for the Elderly)	5, 7, 10, 24
Prune for Better Fruit	2, 3, 4, 30, 31
Prune to Promote Healing	
Wean Children	23, 24, 25
Wood Floors (Installing)	
Write Letters or Contracts	4, 7, 8, 18, 23, 31

Choose the Best Time for Your Activities

When rules for elections refer to "favorable" and "unfavorable" aspects to your Sun or other planets, please refer to the Favorable and Unfavorable Days Tables and Lunar Aspectarian for more information. You'll find instructions beginning on page 133 and the tables beginning on page 141.

The material in this section came from several sources including: *The New A to Z Horoscope Maker and Delineator* by Llewellyn George (Llewellyn, 1999), *Moon Sign Book* (Llewellyn, 1945), and *Electional Astrology* by Vivian Robson (Slingshot Publishing, 2000). Robson's book was originally published in 1937.

Advertise (Internet)

The Moon should be conjunct, sextile, or trine Mercury or Uranus and in the sign of Gemini, Capricorn, or Aquarius.

Advertise (Print)

Write ads on a day favorable to your Sun. The Moon should be conjunct, sextile, or trine Mercury or Venus. Avoid hard aspects to Mars and Saturn. Ad campaigns produce the best results when the Moon is well aspected in Gemini (to enhance communication) or Capricorn (to build business).

Animals

Take home new pets when the day is favorable to your Sun, or when the Moon is trine, sextile, or conjunct Mercury, Venus, or Jupiter, or in the sign of Virgo or Pisces. However, avoid days when the Moon is either square or opposing the Sun, Mars, Saturn, Uranus, Neptune, or Pluto. When selecting a pet, have the Moon well aspected by the planet that rules the animal. Cats are ruled by the Sun, dogs by Mercury, birds by Venus, horses by Jupiter, and fish by Neptune. Buy large animals when the Moon is in Sagittarius or Pisces and making favorable aspects to Jupiter or Mercury. Buy animals smaller than sheep when the Moon is in Virgo with favorable aspects to Mercury or Venus.

Animals (Breed)

Animals are easiest to handle when the Moon is in Taurus, Cancer, Libra, or Pisces, but try to avoid the Full Moon. To encourage healthy births, animals should be mated so births occur when the Moon is increasing in Taurus, Cancer, Pisces, or Libra. Those born during a semi-fruitful sign (Taurus and Capricorn) will produce leaner meat. Libra yields beautiful animals for showing and racing.

Animals (Declaw)

Declaw cats in the dark of the Moon. Avoid the week before and after the Full Moon and the sign of Pisces.

Animals (Neuter or spay)

Have livestock and pets neutered or spayed when the Moon is in Sagittarius, Capricorn, or Pisces, after it has passed through Scorpio, the sign that rules reproductive organs. Avoid the week before and after the Full Moon.

Animals (Sell or buy)

In either buying or selling, it is important to keep the Moon and Mercury free from any aspect to Mars. Aspects to Mars will create discord and increase the likelihood of wrangling over price and quality. The Moon should be passing from the first quarter to full and sextile or trine Venus or Jupiter. When buying racehorses, let the Moon be in an air sign. The Moon should be in air signs when you buy birds. If the birds are to be pets, let the Moon be in good aspect to Venus.

Animals (Train)

Train pets when the Moon is in Virgo or trine to Mercury.

Animals (Train dogs to hunt)

Let the Moon be in Aries in conjunction with Mars, which makes them courageous and quick to learn. But let Jupiter also be in aspect to preserve them from danger in hunting.

Automobiles

When buying an automobile, select a time when the Moon is conjunct, sextile, or trine to Mercury, Saturn, or Uranus and in the sign of Gemini or Capricorn. Avoid times when Mercury is in retrograde motion.

Baking Cakes

Your cakes will have a lighter texture if you see that the Moon is in Gemini, Libra, or Aquarius and in good aspect to Venus or Mercury. If you are decorating a cake or confections are being made, have the Moon placed in Libra.

Beauty Treatments (Massage, etc.)

See that the Moon is in Taurus, Cancer, Leo, Libra, or Aquarius and in favorable aspect to Venus. In the case of plastic surgery, aspects to Mars should be avoided, and the Moon should not be in the sign ruling the part to be operated on.

Borrow (Money or goods)

See that the Moon is not placed between 15 degrees Libra and 15 degrees Scorpio. Let the Moon be waning and in Leo, Scorpio (16 to 30 degrees), Sagittarius, or Pisces. Venus should be in good aspect to the Moon, and the Moon should not be square, opposing, or conjunct either Saturn or Mars.

Brewing

Start brewing during the third or fourth quarter, when the Moon is in Cancer, Scorpio, or Pisces.

Build (Start foundation)

Turning the first sod for the foundation marks the beginning of the building. For best results, excavate the site when the Moon is in the first quarter of a fixed sign and making favorable aspects to Saturn.

Business (Start new)

When starting a business, have the Moon be in Taurus, Virgo, or Capricorn and increasing. The Moon should be sextile or trine Jupiter or Saturn, but avoid oppositions or squares. The planet ruling the business should be well aspected, too.

Buy Goods

Buy during the third quarter, when the Moon is in Taurus for quality or in a mutable sign (Gemini, Sagittarius, Virgo, or Pisces) for savings. Good aspects to Venus or the Sun are desirable. If you are buying for yourself, it is good if the day is favorable for your Sun sign. You may also apply rules for buying specific items.

Canning

Can fruits and vegetables when the Moon is in either the third or fourth quarter and in the water sign Cancer or Pisces. Preserves and jellies use the same quarters and the signs Cancer, Pisces, or Taurus.

Clothing

Buy clothing on a day that is favorable for your Sun sign and when Venus or Mercury is well aspected. Avoid aspects to Mars and Saturn. Buy your clothing when the Moon is in Taurus if you want to remain satisfied. Do not buy clothing or jewelry when the Moon is in Scorpio or Aries. See that the Moon is sextile or trine the Sun during the first or second quarters.

Collections

Try to make collections on days when your Sun is well aspected. Avoid days when the Moon is opposing or square Mars or Saturn. If possible, the Moon should be in a cardinal sign (Aries, Cancer, Libra, or Capricorn). It is more difficult to collect when the Moon is in Taurus or Scorpio.

Concrete

Pour concrete when the Moon is in the third quarter of the fixed sign Taurus, Leo, or Aquarius.

Construction (Begin new)

The Moon should be sextile or trine Jupiter. According to Hermes, no building should be begun when the Moon is in Scorpio or Pisces. The best time to begin building is when the Moon is in Aquarius.

Consultants (Work with)

The Moon should be conjunct, sextile, or trine Mercury or Jupiter.

Contracts (Bid on)

The Moon should be in Gemini or Capricorn and either the Moon or Mercury should be conjunct, sextile, or trine Jupiter.

Copyrights/Patents

The Moon should be conjunct, trine, or sextile either Mercury or Jupiter.

Coronations and Installations

Let the Moon be in Leo and in favorable aspect to Venus, Jupiter, or Mercury. The Moon should be applying to these planets.

Cultivate

Cultivate when the Moon is in a barren sign and waning, ideally the fourth quarter in Aries, Gemini, Leo, Virgo, or Aquarius. The third quarter in the sign of Sagittarius will also work.

Cut Timber

Timber cut during the waning Moon does not become worm-eaten; it will season well and not warp, decay, or snap during burning. Cut when the Moon is in Taurus, Gemini, Virgo, or Capricorn—especially in August. Avoid the water signs. Look for favorable aspects to Mars.

Decorating or Home Repairs

Have the Moon waxing and in the sign of Libra, Gemini, or Aquarius. Avoid squares or oppositions to either Mars or Saturn. Venus in good aspect to Mars or Saturn is beneficial.

Demolition

Let the waning Moon be in Leo, Sagittarius, or Aries.

Dental and Dentists

Visit the dentist when the Moon is in Virgo, or pick a day marked favorable for your Sun sign. Mars should be marked sextile, conjunct, or trine; avoid squares or oppositions to Saturn, Uranus, or Jupiter.

Teeth are best removed when the Moon is in Gemini, Virgo, Sagittarius, or Pisces and during the first or second quarter. Avoid the Full Moon! The day should be favorable for your lunar cycle, and Mars and Saturn should be marked conjunct, trine, or sextile. Fillings should be done in the third or fourth quarters in the sign of Taurus, Leo, Scorpio, or Pisces. The same applies for dentures.

Dressmaking

William Lilly wrote in 1676: "Make no new clothes, or first put them on when the Moon is in Scorpio or afflicted by Mars, for they will be apt to be torn and quickly worn out." Design, repair, and sew clothes in the first and second quarters of Taurus, Leo, or Libra on a day marked favorable for your Sun sign. Venus, Jupiter, and Mercury should be favorably aspected, but avoid hard aspects to Mars or Saturn.

Egg-setting

Eggs should be set so chicks will hatch during fruitful signs. To set eggs, subtract the number of days given for incubation or gestation from the fruitful dates. Chickens incubate in twenty-one days, turkeys and geese in twenty-eight days.

A freshly laid egg loses quality rapidly if it is not handled properly. Use plenty of clean litter in the nests to reduce the number of dirty or cracked eggs. Gather eggs daily in mild weather and at least two times daily in hot or cold weather. The eggs should be placed in a cooler immediately after gathering and stored at 50 to 55 degrees Fahrenheit. Do not store eggs with foods or products that give off pungent odors since eggs may absorb the odors.

Eggs saved for hatching purposes should not be washed. Only clean and slightly soiled eggs should be saved for hatching. Dirty eggs should not be incubated. Eggs should be stored in a cool place with the large ends up. It is not advisable to store the eggs longer than one week before setting them in an incubator.

Electricity and Gas (Install)

The Moon should be in a fire sign, and there should be no squares, oppositions, or conjunctions with Uranus (ruler of electricity), Neptune (ruler of gas), Saturn, or Mars. Hard aspects to Mars can cause fires.

Electronics (Buying)

Choose a day when the Moon is in an air sign (Gemini, Libra, Aquarius) and well aspected by Mercury and/or Uranus when buying electronics.

Electronics (Repair)

The Moon should be sextile or trine Mars or Uranus and in a fixed sign (Taurus, Leo, Scorpio, Aquarius).

Entertain Friends

Let the Moon be in Leo or Libra and making good aspects to Venus. Avoid squares or oppositions to either Mars or Saturn by the Moon or Venus.

Eyes and Eyeglasses

Have your eyes tested and glasses fitted on a day marked favorable for your Sun sign, and on a day that falls during your favorable lunar cycle. Mars should not be in aspect with the Moon. The same applies for any treatment of the eyes, which should also be started during the Moon's first or second quarter.

Fence Posts

Set posts when the Moon is in the third or fourth quarter of the fixed sign Taurus or Leo.

Fertilize and Compost

Fertilize when the Moon is in a fruitful sign (Cancer, Scorpio, Pisces). Organic fertilizers are best when the Moon is waning. Use chemical fertilizers when the Moon is waxing. Start compost when the Moon is in the fourth quarter in a water sign.

Find Hidden Treasure

Let the Moon be in good aspect to Jupiter or Venus. If you erect a horoscope for this election, place the Moon in the Fourth House.

Find Lost Articles

Search for lost articles during the first quarter and when your Sun sign is marked favorable. Also check to see that the planet ruling the lost item is trine, sextile, or conjunct the Moon. The Moon rules household utensils; Mercury rules letters and books; and Venus rules clothing, jewelry, and money.

Fishing

During the summer months, the best time of the day to fish is from sunrise to three hours after and from two hours before sunset until one hour after. Fish do not bite in cooler months until the air is warm, from noon to 3 pm. Warm, cloudy days are good. The most favorable winds are from the south and southwest. Easterly winds are unfavorable. The best days of the month for fishing are when the Moon changes quarters, especially if the change occurs on a day when the Moon is in a water sign (Cancer, Scorpio, Pisces). The best period in any month is the day after the Full Moon.

Friendship

The need for friendship is greater when the Moon is in Aquarius or when Uranus aspects the Moon. Friendship prospers when Venus or Uranus is trine, sextile, or conjunct the Moon. The Moon in Gemini facilitates the chance meeting of acquaintances and friends.

Grafting or Budding

Grafting is the process of introducing new varieties of fruit on less desirable trees. For this process you should use the increasing phase of the Moon in fruitful signs such as Cancer, Scorpio, or Pisces. Capricorn may be used, too. Cut your grafts while trees are dormant, from December to March. Keep them in a cool, dark place, not too dry or too damp. Do the grafting before the sap starts to flow and while the Moon is waxing, preferably while it is in Cancer, Scorpio, or Pisces. The type of plant should determine both cutting and planting times.

Habit (Breaking)

To end an undesirable habit, and this applies to ending everything from a bad relationship to smoking, start on a day when the Moon is in the fourth quarter and in the barren sign of Gemini, Leo, or

Aquarius. Aries, Virgo, and Capricorn may be suitable as well, depending on the habit you want to be rid of. Make sure that your lunar cycle is favorable. Avoid lunar aspects to Mars or Jupiter. However, favorable aspects to Pluto are helpful.

Haircuts

Cut hair when the Moon is in Gemini, Sagittarius, Pisces, Taurus, or Capricorn, but not in Virgo. Look for favorable aspects to Venus. For faster growth, cut hair when the Moon is increasing in Cancer or Pisces. To make hair grow thicker, cut when the Moon is full in the signs of Taurus, Cancer, or Leo. If you want your hair to grow more slowly, have the Moon be decreasing in Aries, Gemini, or Virgo, and have the Moon square or opposing Saturn.

Permanents, straightening, and hair coloring will take well if the Moon is in Taurus or Leo and trine or sextile Venus. Avoid hair treatments if Mars is marked as square or in opposition, especially if heat is to be used. For permanents, a trine to Jupiter is helpful. The Moon also should be in the first quarter. Check the lunar cycle for a favorable day in relation to your Sun sign.

Harvest Crops

Harvest root crops when the Moon is in a dry sign (Aries, Leo, Sagittarius, Gemini, Aquarius) and waning. Harvest grain for storage just after the Full Moon, avoiding Cancer, Scorpio, or Pisces. Harvest in the third and fourth quarters in dry signs. Dry crops in the third quarter in fire signs.

Health

A diagnosis is more likely to be successful when the Moon is in Aries, Cancer, Libra, or Capricorn and less so when in Gemini, Sagittarius, Pisces, or Virgo. Begin a recuperation program when the Moon is in a cardinal or fixed sign and the day is favorable to your Sun sign. Enter hospitals at these times, too. For surgery, see "Surgical Procedures." Buy medicines when the Moon is in Virgo or Scorpio.

Home (Buy new)

If you desire a permanent home, buy when the New Moon is in a fixed sign—Taurus or Leo—for example. Each sign will affect your decision in a different way. A house bought when the Moon is in Taurus is likely to be more practical and have a country look—right down to the split-rail fence. A house purchased when the Moon is in Leo will more likely be a real showplace.

If you're buying for speculation and a quick turnover, be certain that the Moon is in a cardinal sign (Aries, Cancer, Libra, Capricorn). Avoid buying when the Moon is in a fixed sign (Leo, Scorpio, Aquarius, Taurus).

Home (Make repairs)

In all repairs, avoid squares, oppositions, or conjunctions to the planet ruling the place or thing to be repaired. For example, bathrooms are ruled by Scorpio and Cancer. You would not want to start a project in those rooms when the Moon or Pluto is receiving hard aspects. The front entrance, hall, dining room, and porch are ruled by the Sun. So you would want to avoid times when Saturn or Mars are square, opposing, or conjunct the Sun. Also, let the Moon be waxing.

Home (Sell)

Make a strong effort to list your property for sale when the Sun is marked favorable in your sign and in good aspect to Jupiter. Avoid adverse aspects to as many planets as possible.

Home Furnishings (Buy new)

Saturn days (Saturday) are good for buying, and Jupiter days (Thursday) are good for selling. Items bought on days when Saturn is well aspected tend to wear longer and purchases tend to be more conservative.

Job (Start new)

Jupiter and Venus should be sextile, trine, or conjunct the Moon. A day when your Sun is receiving favorable aspects is preferred.

Legal Matters

Good Moon-Jupiter aspects improve the outcome in legal decisions. To gain damages through a lawsuit, begin the process during the increasing Moon. To avoid paying damages, a court date during the decreasing Moon is desirable. Good Moon-Sun aspects strengthen your chance of success. A well-aspected Moon in Cancer or Leo, making good aspects to the Sun, brings the best results in custody cases. In divorce cases, a favorable Moon-Venus aspect is best.

Loan (Ask for)

A first and second quarter phase favors the lender, the third and fourth quarters favor the borrower. Good aspects of Jupiter and Venus to the Moon are favorable to both, as is having the Moon in Leo or Taurus.

Machinery, Appliances, or Tools (Buy)

Tools, machinery, and other implements should be bought on days when your lunar cycle is favorable and when Mars and Uranus are trine, sextile, or conjunct the Moon. Any quarter of the Moon is suitable. When buying gas or electrical appliances, the Moon should be in Aquarius.

Make a Will

Let the Moon be in a fixed sign (Taurus, Leo, Scorpio, or Aquarius) to ensure permanence. If the Moon is in a cardinal sign (Aries, Cancer, Libra, or Capricorn), the will could be altered. Let the Moon be waxing—increasing in light—and in good aspect to Saturn, Venus, or Mercury. In case the will is made in an emergency during illness and the Moon is slow in motion, void-of-course, combust, or under the Sun's beams, the testator will die and the will remain unaltered. There is some danger that it will be lost or stolen, however.

Marriage

The best time for marriage to take place is when the Moon is increasing, but not yet full. Good signs for the Moon to be in are Taurus, Cancer, Leo, or Libra.

The Moon in Taurus produces the most steadfast marriages, but if the partners later want to separate, they may have a difficult time. Make sure that the Moon is well aspected, especially to Venus or Jupiter. Avoid aspects to Mars, Uranus, or Pluto and the signs Aries, Gemini, Virgo, Scorpio, or Aquarius.

The values of the signs are as follows:

- Aries is not favored for marriage
- Taurus from 0 to 19 degrees is good, the remaining degrees are less favorable
- Cancer is unfavorable unless you are marrying a widow
- Leo is favored, but it may cause one party to deceive the other as to his or her money or possessions
- Virgo is not favored except when marrying a widow
- Libra is good for engagements but not for marriage
- Scorpio from 0 to 15 degrees is good, but the last 15 degrees are entirely unfortunate. The woman may be fickle, envious, and quarrelsome
- Sagittarius is neutral
- Capricorn, from 0 to 10 degrees, is difficult for marriage; however, the remaining degrees are favorable, especially when marrying a widow
- Aquarius is not favored
- Pisces is favored, although marriage under this sign can incline a woman to chatter a lot

These effects are strongest when the Moon is in the sign. If the Moon and Venus are in a cardinal sign, happiness between the couple may not continue long.

On no account should the Moon apply to Saturn or Mars, even by good aspect.

Medical Treatment for the Eyes

Let the Moon be increasing in light and motion and making favorable aspects to Venus or Jupiter and be unaspected by Mars. Keep the Moon out of Taurus, Capricorn, or Virgo. If an aspect between the Moon and Mars is unavoidable, let it be separating.

Medical Treatment for the Head

If possible, have Mars and Saturn free of hard aspects. Let the Moon be in Aries or Taurus, decreasing in light, in conjunction or aspect with Venus or Jupiter and free of hard aspects. The Sun should not be in any aspect to the Moon.

Medical Treatment for the Nose

Let the Moon be in Cancer, Leo, or Virgo and not aspecting Mars or Saturn and also not in conjunction with a retrograde or weak planet.

Mining

Saturn rules mining. Begin work when Saturn is marked conjunct, trine, or sextile. Mine for gold when the Sun is marked conjunct, trine, or sextile. Mercury rules quicksilver, Venus rules copper, Jupiter rules tin, Saturn rules lead and coal, Uranus rules radioactive elements, Neptune rules oil, the Moon rules water. Mine for these items when the ruling planet is marked conjunct, trine, or sextile.

2009 © Andy Dean. Image from BigStockPhoto.com

Move to New Home

If you have a choice, and sometimes we don't, make sure that Mars is not aspecting the Moon. Move on a day favorable to your Sun sign or when the Moon is conjunct, sextile, or trine the Sun.

Mow Lawn

Mow in the first and second quarters (waxing phase) to increase growth and lushness, and in the third and fourth quarters (waning phase) to decrease growth.

Negotiate

When you are choosing a time to negotiate, consider what the meeting is about and what you want to have happen. If it is agreement or compromise between two parties that you desire, have the Moon be in the sign of Libra. When you are making contracts, it is best to have the Moon in the same element. For example, if your concern is communication, then elect a time when the Moon is in an air sign. If, on the other hand, your concern is about possessions, an earth sign would be more appropriate. Fixed signs are unfavorable, with the exception of Leo; so are cardinal signs, except for Capricorn. If you are negotiating the end of something, use the rules that apply to ending habits.

Occupational Training

When you begin training, see that your lunar cycle is favorable that day and that the planet ruling your occupation is marked conjunct or trine.

Paint

Paint buildings during the waning Libra or Aquarius Moon. If the weather is hot, paint when the Moon is in Taurus. If the weather is cold, paint when the Moon is in Leo. Schedule the painting to start in the fourth quarter as the wood is drier and paint will penetrate wood better. Avoid painting around the New Moon, though, as the wood is likely to be damp, making the paint subject to scalding when hot weather hits it. If the temperature is below 70 degrees Fahrenheit, it is not advisable to paint while the Moon is in Cancer, Scorpio, or Pisces as the paint is apt to creep, check, or run.

Party (Host or attend)

A party timed so the Moon is in Gemini, Leo, Libra, or Sagittarius, with good aspects to Venus and Jupiter, will be fun and well attended. There should be no aspects between the Moon and Mars or Saturn.

Pawn

Do not pawn any article when Jupiter is receiving a square or opposition from Saturn or Mars or when Jupiter is within 17 degrees of the Sun, for you will have little chance to redeem the items.

Pick Mushrooms

Mushrooms, one of the most promising traditional medicines in the world, should be gathered at the Full Moon.

Plant

Root crops, like carrots and potatoes, are best if planted in the sign Taurus or Capricorn. Beans, peas, tomatoes, peppers, and other fruit-bearing plants are best if planted in a sign that supports seed growth. Leaf plants, like lettuce, broccoli, or cauliflower, are best planted when the Moon is in a water sign.

It is recommended that you transplant during a decreasing Moon, when forces are streaming into the lower part of the plant. This helps root growth.

Promotion (Ask for)

Choose a day favorable to your Sun sign. Mercury should be marked conjunct, trine, or sextile. Avoid days when Mars or Saturn is aspected.

Prune

Prune during the third and fourth quarter of a Scorpio Moon to retard growth and to promote better fruit. Prune when the Moon is in cardinal Capricorn to promote healing.

Reconcile with People

If the reconciliation be with a woman, let Venus be strong and well aspected. If elders or superiors are involved, see that Saturn is receiving good aspects; if the reconciliation is between young people or between an older and younger person, see that Mercury is well aspected.

Romance

There is less control of when a romance starts, but romances begun under an increasing Moon are more likely to be permanent or satisfying, while those begun during the decreasing Moon tend to transform the participants. The tone of the relationship can be guessed from the sign the Moon is in. Romances begun with the Moon in Aries may be impulsive. Those begun in Capricorn will take greater effort to bring to a desirable conclusion, but they may be very rewarding. Good aspects between the Moon and Venus will have a positive influence on the relationship. Avoid unfavorable aspects to Mars, Uranus, and Pluto. A decreasing Moon, particularly the fourth quarter, facilitates ending a relationship and causes the least pain.

Roof a Building

Begin roofing a building during the third or fourth quarter, when the Moon is in Aries or Aquarius. Shingles laid during the New Moon have a tendency to curl at the edges.

Sauerkraut

The best-tasting sauerkraut is made just after the Full Moon in the fruitful signs of Cancer, Scorpio, or Pisces.

Select a Child's Sex

Count from the last day of menstruation to the first day of the next cycle and divide the interval between the two dates in half. Pregnancy in the first half produces females, but copulation should take place with the Moon in a feminine sign. Pregnancy in the latter half, up to three days before the beginning of menstruation, produces males, but copulation should take place with the Moon in a masculine sign. The three-day period before the next period again produces females.

Sell or Canvass

Begin these activities during a day favorable to your Sun sign. Otherwise, sell on days when Jupiter, Mercury, or Mars is trine, sextile, or conjunct the Moon. Avoid days when Saturn is square or opposing the Moon, for that always hinders business and causes discord. If the Moon is passing from the first quarter to full, it is best to have the Moon swift in motion and in good aspect with Venus and/or Jupiter.

Sign Papers

Sign contracts or agreements when the Moon is increasing in a fruitful sign and on a day when the Moon is making favorable aspects to Mercury. Avoid days when Mars, Saturn, or Neptune are square or opposite the Moon.

Spray and Weed

Spray pests and weeds during the fourth quarter when the Moon is in the barren sign Leo or Aquarius and making favorable aspects to Pluto. Weed during a waning Moon in a barren sign.

Staff (Fire)

Have the Moon in the third or fourth quarter, but not full. The Moon should not be square any planets.

Staff (Hire)

The Moon should be in the first or second quarter, and preferably in the sign of Gemini or Virgo. The Moon should be conjunct, trine, or sextile Mercury or Jupiter.

Stocks (Buy)

The Moon should be in Taurus or Capricorn, and there should be a sextile or trine to Jupiter or Saturn.

Surgical Procedures

Blood flow, like ocean tides, appears to be related to Moon phases. To reduce hemorrhage after a surgery, schedule it within one week before or after a New Moon. Schedule surgery to occur during the increase of the Moon if possible, as wounds heal better and vitality is greater than during the decrease of the Moon. Avoid surgery within one week before or after the Full Moon. Select a date when the Moon is past the sign governing the part of the body involved in the operation. For example, abdominal operations should be done when the Moon is in Sagittarius, Capricorn, or Aquarius. The further removed the Moon sign is from the sign ruling the afflicted part of the body, the better.

For successful operations, avoid times when the Moon is applying to any aspect of Mars. (This tends to promote inflammation and complications.) See the Lunar Aspectarian on odd pages 141–163 to find days with negative Mars aspects and positive Venus and Jupiter aspects. Never operate with the Moon in the same sign as a person's Sun sign or Ascendant. Let the Moon be in a fixed sign and avoid square or opposing aspects. The Moon should not be void-of-course. Cosmetic surgery should be done in the increase of the Moon, when the Moon is not square or in opposition to Mars. Avoid days when the Moon is square or opposing Saturn or the Sun.

Travel (Air)

Start long trips when the Moon is making favorable aspects to the Sun. For enjoyment, aspects to Jupiter are preferable; for visiting, look for favorable aspects to Mercury. To prevent accidents, avoid squares or oppositions to Mars, Saturn, Uranus, or Pluto. Choose a day when the Moon is in Sagittarius or Gemini and well aspected to Mercury, Jupiter, or Uranus. Avoid adverse aspects of Mars, Saturn, or Uranus.

Visit

On setting out to visit a person, let the Moon be in aspect with any retrograde planet, for this ensures that the person you're visiting will be at home. If you desire to stay a long time in a place, let the Moon be in good aspect to Saturn. If you desire to leave the place quickly, let the Moon be in a cardinal sign.

Wean Children

To wean a child successfully, do so when the Moon is in Sagittarius, Capricorn, Aquarius, or Pisces—signs that do not rule vital human organs. By observing this astrological rule, much trouble for parents and child may be avoided.

Weight (Reduce)

If you want to lose weight, the best time to get started is when the Moon is in the third or fourth quarter and in the barren sign of Virgo. Review the section on How to Use the Moon Tables and Lunar Aspectarian beginning on page 141 to help you select a date that is favorable to begin your weight-loss program.

Wine and Drink Other Than Beer

Start brewing when the Moon is in Pisces or Taurus. Sextiles or trines to Venus are favorable, but avoid aspects to Mars or Saturn.

Write

Write for pleasure or publication when the Moon is in Gemini. Mercury should be making favorable aspects to Uranus and Neptune.

How to Use the Moon Tables and Lunar Aspectarian

Timing activities is one of the most important things you can do to ensure success. In many Eastern countries, timing by the planets is so important that practically no event takes place without first setting up a chart for it. Weddings have occurred in the middle of the night because the influences were best then. You may not want to take it that far, but you can still make use of the influences of the Moon whenever possible. It's easy and it works!

The *Moon Sign Book* has information to help you plan just about any activity: weddings, fishing, making purchases, cutting your hair, traveling, and more. We provide the guidelines you need to pick the best day out of the several from which you have to choose. The Moon Tables are the *Moon Sign Book's* primary method for choosing dates. Following are instructions, examples, and directions on how to read the Moon Tables. More advanced information on using the tables containing the Lunar Aspectarian and favorable and unfavorable days (found on odd-numbered pages opposite the Moon Tables), Moon void-of-course and retrograde information to choose the dates best for you is also included.

The Five Basic Steps

Step 1: Directions for Choosing Dates
Look up the directions for choosing dates for the activity that you wish to begin, then go to step 2.

Step 2: Check the Moon Tables
You'll find two tables for each month of the year beginning on page 140. The Moon Tables (on the left-hand pages) include the day, date, and sign the Moon is in; the element and nature of the sign; the Moon's phase; and when it changes sign or phase. If there is a time listed after a date, that time is the time when the Moon moves into that zodiac sign. Until then, the Moon is considered to be in the sign for the previous day.

The abbreviation Full signifies Full Moon and New signifies New Moon. The times listed with dates indicate when the Moon changes sign. The times listed after the phase indicate when the Moon changes phase.

Turn to the month you would like to begin your activity. You will be using the Moon's sign and phase information most often when you begin choosing your own dates. Use the Time Zone Map on page 168 and the Time Zone Conversions table on page 169 to convert time to your own time zone.

When you find dates that meet the criteria for the correct Moon phase and sign for your activity, you may have completed the process. For certain simple activities, such as getting a hair-cut, the phase and sign information is all that is needed. If the directions for your activity include information on certain lunar aspects, however, you should consult the Lunar Aspectarian. An example of this would be if the directions told you not to perform a certain activity when the Moon is square (Q) Jupiter.

Step 3: Check the Lunar Aspectarian

On the pages opposite the Moon Tables you will find tables containing the Lunar Aspectarian and Favorable and Unfavorable Days. The Lunar Aspectarian gives the aspects (or angles) of the Moon to other planets. Some aspects are favorable, while others are not. To use the Lunar Aspectarian, find the planet that the directions list as favorable for your activity, and run down the column to the date desired. For example, you should avoid aspects to Mars if you are planning surgery. So you would look for Mars across the top and then run down that column looking for days where there are no aspects to Mars (as signified by empty boxes). If you want to find a favorable aspect (sextile (X) or trine (T)) to Mercury, run your finger down the column under Mercury until you find an X or T. Adverse aspects to planets are squares (Q) or oppositions (O). A conjunction (C) is sometimes beneficial, sometimes not, depending on the activity or planets involved.

Step 4: Favorable and Unfavorable Days

The tables listing favorable and unfavorable days are helpful when you want to choose your personal best dates because your Sun sign is taken into consideration. The twelve Sun signs are listed on the right side of the tables. Once you have determined which days meet your criteria for phase, sign, and aspects, you can determine whether or not those days are positive for you by checking the favorable and unfavorable days for your Sun sign.

To find out if a day is positive for you, find your Sun sign and then look down the column. If it is marked F, it is very favorable. The Moon is in the same sign as your Sun on a favorable day. If it is marked f, it is slightly favorable; U is very unfavorable; and u means slightly unfavorable. A day marked very unfavorable (U) indicates that the Moon is in the sign opposing your Sun.

Once you have selected good dates for the activity you are about to begin, you can go straight to "Using What You've Learned," beginning on the next page. To learn how to fine-tune your selections even further, read on.

Step 5: Void-of-Course Moon and Retrogrades

This last step is perhaps the most advanced portion of the procedure. It is generally considered poor timing to make decisions, sign important papers, or start special activities during a Moon void-of-course period or during a Mercury retrograde. Once you have chosen the best date for your activity based on steps one through four, you can check the Void-of-Course tables, beginning on page 80, to find out if any of the dates you have chosen have void periods.

The Moon is said to be void-of-course after it has made its last aspect to a planet within a particular sign, but before it has moved into the next sign. Put simply, the Moon is "resting" during the void-of-course period, so activities initiated at this time generally don't come to fruition. You will notice that there are many void periods during the year, and it is nearly impossible to avoid all of them. Some people choose to ignore these altogether and do not take them into consideration when planning activities.

Next, you can check the Retrograde Planets tables on page 164 to see what planets are retrograde during your chosen date(s).

A planet is said to be retrograde when it appears to move backward in the sky as viewed from the Earth. Generally, the farther a planet is away from the Sun, the longer it can stay retrograde. Some planets will retrograde for several months at a time. Avoiding retrogrades is not as important in lunar planning as avoiding the Moon void-of-course, with the exception of the planet Mercury.

Mercury rules thought and communication, so it is advisable not to sign important papers, initiate important business or legal work, or make crucial decisions during these times. As with the Moon void-of-course, it is difficult to avoid all planetary retrogrades when beginning events, and you may choose to ignore this step of the process. Following are some examples using some or all of the steps outlined above.

Using What You've Learned

Let's say it's a new year and you want to have your hair cut. It's thin and you would like it to look fuller, so you find the directions for hair care and you see that for thicker hair you should cut hair while the Moon is Full and in the sign of Taurus, Cancer, or Leo. You should avoid the Moon in Aries, Gemini, or Virgo. Look at the January Moon Table on page 140. You see that the Full Moon is on January 30 at 1:18 am. The Moon moved into the sign of Leo the previous day and remains in Leo until January 31 at 8:23 am, so January 30–31 meets both the phase and sign criteria.

Let's move on to a more difficult example using the sign and phase of the Moon. You want to buy a permanent home. After checking the instructions for purchasing a house: "Home (Buy new)" on page 122, you see that you should buy a home when the Moon is in Taurus, Cancer, or Leo. You need to get a loan, so you should also look under "Loan (Ask for)" on page 123. Here it says that the third and fourth quarters favor the borrower (you). You are going to buy the house in October so go to page 158. The Moon is in the third quarter October 22–30. The Moon is in Cancer from 3:14 pm on October 27 until October 29 at 8:39 pm. The best days for obtaining a loan would be October 27–29, while the Moon is in Cancer.

Just match up the best sign and phase (quarter) to come up with the best date. With all activities, be sure to check the favorable and unfavorable days for your Sun sign in the table adjoining the Lunar Aspectarian. If there is a choice between several dates, pick the one most favorable for you. Because buying a home is an important business decision, you may also wish to see if the Moon is void or if Mercury is retrograde during these dates.

Now let's look at an example that uses signs, phases, and aspects. Our example is starting new home construction. We will use the month of April. Look under "Build (Start foundation)" on page 114 and you'll see that the Moon should be in the first quarter of Taurus or Leo. You should select a time when the Moon is not making unfavorable aspects to Saturn. (Conjunctions are usually considered good if they are not to Mars, Saturn, or Neptune.) Look in the April Moon Table. You will see that the Moon is in the first quarter April 14–21. The Moon is in Taurus from 6:55 pm on April 14 until April 17 at 2:08 am. Now, look to the April Lunar Aspectarian. We see that there are no squares or oppositions to Saturn between April 14–17. These are good dates to start a foundation.

A Note About Time and Time Zones

All tables in the *Moon Sign Book* use Eastern Time. You must calculate the difference between your time zone and the Eastern Time Zone. Please refer to the Time Zone Conversions chart on 169 for help with time conversions. The sign the Moon is in at midnight is the sign shown in the Aspectarian and Favorable and Unfavorable Days tables.

How Does the Time Matter?

Due to the three-hour time difference between the East and West Coasts of the United States, those of you living on the East Coast may be, for example, under the influence of a Virgo Moon, while those of you living on the West Coast will still have a Leo Moon influence.

We follow a commonly held belief among astrologers: whatever sign the Moon is in at the start of a day—12:00 am Eastern

Time—is considered the dominant influence of the day. That sign is indicated in the Moon Tables. If the date you select for an activity shows the Moon changing signs, you can decide how important the sign change may be for your specific election and adjust your election date and time accordingly.

Use Common Sense

Some activities depend on outside factors. Obviously, you can't go out and plant when there is a foot of snow on the ground. You should adjust to the conditions at hand. If the weather was bad during the first quarter, when it was best to plant crops, do it during the second quarter while the Moon is in a fruitful sign. If the Moon is not in a fruitful sign during the first or second quarter, choose a day when it is in a semi-fruitful sign. The best advice is to choose either the sign or phase that is most favorable, when the two don't coincide.

To Summarize

First, look up the activity under the proper heading, then look for the information given in the tables. Choose the best date considering the number of positive factors in effect. If most of the dates are favorable, there is no problem choosing the one that will fit your schedule. However, if there aren't any really good dates, pick the ones with the least number of negative influences. Please keep in mind that the information found here applies in the broadest sense to the events you want to plan or are considering. To be the most effective, when you use electional astrology, you should also consider your own birth chart in relation to a chart drawn for the time or times you have under consideration. The best advice we can offer you is: read the entire introduction to each section.

January Moon Table

Date	Sign	Element	Nature	Phase
1 Fri 9:41 pm	Leo	Fire	Barren	3rd
2 Sat	Leo	Fire	Barren	3rd
3 Sun 9:52 pm	Virgo	Earth	Barren	3rd
4 Mon	Virgo	Earth	Barren	3rd
5 Tue 11:58 pm	Libra	Air	Semi-fruitful	3rd
6 Wed	Libra	Air	Semi-fruitful	3rd
7 Thu	Libra	Air	Semi-fruitful	4th 5:40 am
8 Fri 5:00 am	Scorpio	Water	Fruitful	4th
9 Sat	Scorpio	Water	Fruitful	4th
10 Sun 1:10 pm	Sagittarius	Fire	Barren	4th
11 Mon	Sagittarius	Fire	Barren	4th
12 Tue 11:54 pm	Capricorn	Earth	Semi-fruitful	4th
13 Wed	Capricorn	Earth	Semi-fruitful	4th
14 Thu	Capricorn	Earth	Semi-fruitful	4th
15 Fri 12:17 pm	Aquarius	Air	Barren	New 2:11 am
16 Sat	Aquarius	Air	Barren	1st
17 Sun	Aquarius	Air	Barren	1st
18 Mon 1:17 am	Pisces	Water	Fruitful	1st
19 Tue	Pisces	Water	Fruitful	1st
20 Wed 1:36 pm	Aries	Fire	Barren	1st
21 Thu	Aries	Fire	Barren	1st
22 Fri 11:39 pm	Taurus	Earth	Semi-fruitful	1st
23 Sat	Taurus	Earth	Semi-fruitful	2nd 5:53 am
24 Sun	Taurus	Earth	Semi-fruitful	2nd
25 Mon 6:11 am	Gemini	Air	Barren	2nd
26 Tue	Gemini	Air	Barren	2nd
27 Wed 9:01 am	Cancer	Water	Fruitful	2nd
28 Thu	Cancer	Water	Fruitful	2nd
29 Fri 9:10 am	Leo	Fire	Barren	2nd
30 Sat	Leo	Fire	Barren	Full 1:18 am
31 Sun 8:23 am	Virgo	Earth	Barren	3rd

January Aspectarian/Favorable & Unfavorable Days

Date	Sun	Mercury	Venus	Mars	Jupiter	Saturn	Uranus	Neptune	Pluto	Aries	Taurus	Gemini	Cancer	Leo	Virgo	Libra	Scorpio	Sagittarius	Capricorn	Aquarius	Pisces
1		O						T		u	f		F		f	u	f		U		f
2						X				f	u	f		F		f	u	f		U	
3			C		O			O		f	u	f		F		f	u	f		U	
4	T	T	T						T		f	u	f		F		f	u	f		U
5						O					f	u	f		F		f	u	f		U
6		Q			C				Q	U		f	u	f		F		f	u	f	
7	Q		Q	X				T		U		f	u	f		F		f	u	f	
8		X			T				X	U		f	u	f		F		f	u	f	
9	X		X	Q							U		f	u	f		F		f	u	f
10					Q	X	T	Q			U		f	u	f		F		f	u	f
11				T						f		U		f	u	f		F		f	u
12					X		Q	X		f		U		f	u	f		F		f	u
13		C			Q				C	u	f		U		f	u	f		F		f
14							X			u	f		U		f	u	f		F		f
15	C		C			T				u	f		U		f	u	f		F		f
16			O							f	u	f		U		f	u	f		F	
17								C		f	u	f		U		f	u	f		F	
18		X			C				X		f	u	f		U		f	u	f		F
19											f	u	f		U		f	u	f		F
20	X		X		O	C		Q			f	u	f		U		f	u	f		F
21		Q		T						F		f	u	f		U		f	u	f	
22								X		F		f	u	f		U		f	u	f	
23	Q	T	Q	Q	X			T		F		f	u	f		U		f	u	f	
24						X	Q				F		f	u	f		U		f	u	f
25	T		T		Q	T					F		f	u	f		U		f	u	f
26				X			Q			f		F		f	u	f		U		f	u
27					T	Q		T	O	f		F		f	u	f		U		f	u
28		O						T		u	f		F		f	u	f		U		f
29						X				u	f		F		f	u	f		U		f
30	O		O	C						f	u	f		F		f	u	f		U	
31					O			O	T	f	u	f		F		f	u	f		U	

February Moon Table

Date	Sign	Element	Nature	Phase
1 Mon	Virgo	Earth	Barren	3rd
2 Tue 8:42 am	Libra	Air	Semi-fruitful	3rd
3 Wed	Libra	Air	Semi-fruitful	3rd
4 Thu 11:56 am	Scorpio	Water	Fruitful	3rd
5 Fri	Scorpio	Water	Fruitful	4th 6:48 pm
6 Sat 7:04 pm	Sagittarius	Fire	Barren	4th
7 Sun	Sagittarius	Fire	Barren	4th
8 Mon	Sagittarius	Fire	Barren	4th
9 Tue 5:43 am	Capricorn	Earth	Semi-fruitful	4th
10 Wed	Capricorn	Earth	Semi-fruitful	4th
11 Thu 6:24 pm	Aquarius	Air	Barren	4th
12 Fri	Aquarius	Air	Barren	4th
13 Sat	Aquarius	Air	Barren	New 9:51 pm
14 Sun 7:23 am	Pisces	Water	Fruitful	1st
15 Mon	Pisces	Water	Fruitful	1st
16 Tue 7:30 pm	Aries	Fire	Barren	1st
17 Wed	Aries	Fire	Barren	1st
18 Thu	Aries	Fire	Barren	1st
19 Fri 5:55 am	Taurus	Earth	Semi-fruitful	1st
20 Sat	Taurus	Earth	Semi-fruitful	1st
21 Sun 1:47 pm	Gemini	Air	Barren	2nd 7:42 pm
22 Mon	Gemini	Air	Barren	2nd
23 Tue 6:29 pm	Cancer	Water	Fruitful	2nd
24 Wed	Cancer	Water	Fruitful	2nd
25 Thu 8:08 pm	Leo	Fire	Barren	2nd
26 Fri	Leo	Fire	Barren	2nd
27 Sat 7:52 pm	Virgo	Earth	Barren	2nd
28 Sun	Virgo	Earth	Barren	Full 11:38 am

February Aspectarian/Favorable & Unfavorable Days

Date	Sun	Mercury	Venus	Mars	Jupiter	Saturn	Uranus	Neptune	Pluto	Aries	Taurus	Gemini	Cancer	Leo	Virgo	Libra	Scorpio	Sagittarius	Capricorn	Aquarius	Pisces
1		T					O				f	u	f		F		f	u	f		U
2				X		C			Q		f	u	f		F		f	u	f		U
3	T	Q	T							U		f	u	f		F		f	u	f	
4					T			T	X	U		f	u	f		F		f	u	f	
5	Q			Q							U		f	u	f		F		f	u	f
6		X	Q				T	Q			U		f	u	f		F		f	u	f
7				T	Q	X				f		U		f	u	f		F		f	u
8	X		X				Q	X		f		U		f	u	f		F		f	u
9					X	Q			C	f		U		f	u	f		F		f	u
10										u	f		U		f	u	f		F		f
11		C					X			u	f		U		f	u	f		F		f
12			O		T					f	u	f		U		f	u	f		F	
13	C							C		f	u	f		U		f	u	f		F	
14			C		C				X	f	u	f		U		f	u	f		F	
15											f	u	f		U		f	u	f		F
16						C					f	u	f		U		f	u	f		F
17		X		T		O			Q	F		f	u	f		U		f	u	f	
18								X		F		f	u	f		U		f	u	f	
19	X			Q	X				T	F		f	u	f		U		f	u	f	
20		Q	X								F		f	u	f		U		f	u	f
21	Q			X		T	X	Q			F		f	u	f		U		f	u	f
22			Q		Q					f		F		f	u	f		U		f	u
23		T				Q	Q	T		f		F		f	u	f		U		f	u
24	T		T		T				O	u	f		F		f	u	f		U		f
25				C			T			u	f		F		f	u	f		U		f
26					X					f	u	f		F		f	u	f		U	
27		O						O		f	u	f		F		f	u	f		U	
28	O			O				T			f	u	f		F		f	u	f		U

March Moon Table

Date	Sign	Element	Nature	Phase
1 Mon 7:31 pm	Libra	Air	Semi-fruitful	3rd
2 Tue	Libra	Air	Semi-fruitful	3rd
3 Wed 9:11 pm	Scorpio	Water	Fruitful	3rd
4 Thu	Scorpio	Water	Fruitful	3rd
5 Fri	Scorpio	Water	Fruitful	3rd
6 Sat 2:36 am	Sagittarius	Fire	Barren	3rd
7 Sun	Sagittarius	Fire	Barren	4th 10:42 am
8 Mon 12:13 pm	Capricorn	Earth	Semi-fruitful	4th
9 Tue	Capricorn	Earth	Semi-fruitful	4th
10 Wed	Capricorn	Earth	Semi-fruitful	4th
11 Thu 12:42 am	Aquarius	Air	Barren	4th
12 Fri	Aquarius	Air	Barren	4th
13 Sat 1:44 pm	Pisces	Water	Fruitful	4th
14 Sun	Pisces	Water	Fruitful	4th
15 Mon	Pisces	Water	Fruitful	New 5:01 pm
16 Tue 2:32 am	Aries	Fire	Barren	1st
17 Wed	Aries	Fire	Barren	1st
18 Thu 12:29 pm	Taurus	Earth	Semi-fruitful	1st
19 Fri	Taurus	Earth	Semi-fruitful	1st
20 Sat 8:28 pm	Gemini	Air	Barren	1st
21 Sun	Gemini	Air	Barren	1st
22 Mon	Gemini	Air	Barren	1st
23 Tue 2:16 am	Cancer	Water	Fruitful	2nd 7:00 am
24 Wed	Cancer	Water	Fruitful	2nd
25 Thu 5:39 am	Leo	Fire	Barren	2nd
26 Fri	Leo	Fire	Barren	2nd
27 Sat 6:57 am	Virgo	Earth	Barren	2nd
28 Sun	Virgo	Earth	Barren	2nd
29 Mon 7:21 am	Libra	Air	Semi-fruitful	Full 10:25 pm
30 Tue	Libra	Air	Semi-fruitful	3rd
31 Wed 8:41 am	Scorpio	Water	Fruitful	3rd

March Aspectarian/Favorable & Unfavorable Days

Date	Sun	Mercury	Venus	Mars	Jupiter	Saturn	Uranus	Neptune	Pluto	Aries	Taurus	Gemini	Cancer	Leo	Virgo	Libra	Scorpio	Sagittarius	Capricorn	Aquarius	Pisces
1			O	X			O				f	u	f		F		f	u	f		U
2						C			Q	U		f	u	f		F		f	u	f	
3			Q					T		U		f	u	f		F		f	u	f	
4	T	T			T				X		U		f	u	f		F		f	u	f
5			T					T	Q		U		f	u	f		F		f	u	f
6		Q			T	Q	X			f		U		f	u	f		F		f	u
7	Q									f		U		f	u	f		F		f	u
8			Q			Q	Q	X	C	f		U		f	u	f		F		f	u
9		X		X						u	f		U		f	u	f		F		f
10	X						X			u	f		U		f	u	f		F		f
11			X	O	T					f	u	f		U		f	u	f		F	
12										f	u	f		U		f	u	f		F	
13								C		f	u	f		U		f	u	f		F	
14					C				X		f	u	f		U		f	u	f		F
15	C	C				C					f	u	f		U		f	u	f		F
16			T			O			Q	F		f	u	f		U		f	u	f	
17		C								F		f	u	f		U		f	u	f	
18			Q					X	T	F		f	u	f		U		f	u	f	
19				X							F		f	u	f		U		f	u	f
20	X			X	T		X	Q			F		f	u	f		U		f	u	f
21		X		Q						f		F		f	u	f		U		f	u
22			X				Q	T		f		F		f	u	f		U		f	u
23	Q				Q				O	u	f		F		f	u	f		U		f
24		Q	Q	T						u	f		F		f	u	f		U		f
25	T				C		X	T		u	f		F		f	u	f		U		f
26		T	T							f	u	f		F		f	u	f		U	
27								O	T	f	u	f		F		f	u	f		U	
28				O							f	u	f		F		f	u	f		U
29	O			X		C	O		Q		f	u	f		F		f	u	f		U
30										U		f	u	f		F		f	u	f	
31		O		O	Q			T	X	U		f	u	f		F		f	u	f	

April Moon Table

Date	Sign	Element	Nature	Phase
1 Thu	Scorpio	Water	Fruitful	3rd
2 Fri 12:53 pm	Sagittarius	Fire	Barren	3rd
3 Sat	Sagittarius	Fire	Barren	3rd
4 Sun 9:07 pm	Capricorn	Earth	Semi-fruitful	3rd
5 Mon	Capricorn	Earth	Semi-fruitful	3rd
6 Tue	Capricorn	Earth	Semi-fruitful	4th 5:37 am
7 Wed 8:51 am	Aquarius	Air	Barren	4th
8 Thu	Aquarius	Air	Barren	4th
9 Fri 9:48 pm	Pisces	Water	Fruitful	4th
10 Sat	Pisces	Water	Fruitful	4th
11 Sun	Pisces	Water	Fruitful	4th
12 Mon 9:31 am	Aries	Fire	Barren	4th
13 Tue	Aries	Fire	Barren	4th
14 Wed 6:55 pm	Taurus	Earth	Semi-fruitful	New 8:29 am
15 Thu	Taurus	Earth	Semi-fruitful	1st
16 Fri	Taurus	Earth	Semi-fruitful	1st
17 Sat 2:08 am	Gemini	Air	Barren	1st
18 Sun	Gemini	Air	Barren	1st
19 Mon 7:39 am	Cancer	Water	Fruitful	1st
20 Tue	Cancer	Water	Fruitful	1st
21 Wed 11:42 am	Leo	Fire	Barren	2nd 2:20 pm
22 Thu	Leo	Fire	Barren	2nd
23 Fri 2:24 pm	Virgo	Earth	Barren	2nd
24 Sat	Virgo	Earth	Barren	2nd
25 Sun 4:16 pm	Libra	Air	Semi-fruitful	2nd
26 Mon	Libra	Air	Semi-fruitful	2nd
27 Tue 6:28 pm	Scorpio	Water	Fruitful	2nd
28 Wed	Scorpio	Water	Fruitful	Full 8:19 am
29 Thu 10:36 pm	Sagittarius	Fire	Barren	3rd
30 Fri	Sagittarius	Fire	Barren	3rd

April Aspectarian/Favorable & Unfavorable Days

Date	Sun	Mercury	Venus	Mars	Jupiter	Saturn	Uranus	Neptune	Pluto	Aries	Taurus	Gemini	Cancer	Leo	Virgo	Libra	Scorpio	Sagittarius	Capricorn	Aquarius	Pisces
1					T						U		f	u	f		F		f	u	f
2				T		X	T	Q			U		f	u	f		F		f	u	f
3	T				Q					f		U		f	u	f		F		f	u
4					Q	Q	X			f		U		f	u	f		F		f	u
5		T	T						C	u	f		U		f	u	f		F		f
6	Q				X					u	f		U		f	u	f		F		f
7		Q		O		T	X			u	f		U		f	u	f		F		f
8	X		Q							f	u	f		U		f	u	f		F	
9								C		f	u	f		U		f	u	f		F	
10		X	X						X		f	u	f		U		f	u	f		F
11					C						f	u	f		U		f	u	f		F
12			T			O	C		Q		f	u	f		U		f	u	f		F
13										F		f	u	f		U		f	u	f	
14	C							X		F		f	u	f		U		f	u	f	
15		C		Q					T	F		f	u	f		U		f	u	f	
16			C		X		X	Q		F		f	u	f		U		f	u	f	
17			X			T					F		f	u	f		U		f	u	f
18					Q						F		f	u	f		U		f	u	f
19	X					Q	Q	T	O		F		f	u	f		U		f	u	f
20		X			T					u	f		F		f	u	f		U		f
21	Q		X			X	T			u	f		F		f	u	f		U		f
22		Q	C							f	u	f		F		f	u	f		U	
23	T		Q					O	T	f	u	f		F		f	u	f		U	
24		T									f	u	f		F		f	u	f		U
25			T		O	C	O				f	u	f		F		f	u	f		U
26				X					Q	U		f	u	f		F		f	u	f	
27								T		U		f	u	f		F		f	u	f	
28	O	O		Q					X		U		f	u	f		F		f	u	f
29						T	X	T	Q		U		f	u	f		F		f	u	f
30			O	T						f		U		f	u	f		F		f	u

May Moon Table

Date	Sign	Element	Nature	Phase
1 Sat	Sagittarius	Fire	Barren	3rd
2 Sun 6:00 am	Capricorn	Earth	Semi-fruitful	3rd
3 Mon	Capricorn	Earth	Semi-fruitful	3rd
4 Tue 4:52 pm	Aquarius	Air	Barren	3rd
5 Wed	Aquarius	Air	Barren	3rd
6 Thu	Aquarius	Air	Barren	4th 12:15 am
7 Fri 5:34 am	Pisces	Water	Fruitful	4th
8 Sat	Pisces	Water	Fruitful	4th
9 Sun 5:29 pm	Aries	Fire	Barren	4th
10 Mon	Aries	Fire	Barren	4th
11 Tue	Aries	Fire	Barren	4th
12 Wed 2:48 am	Taurus	Earth	Semi-fruitful	4th
13 Thu	Taurus	Earth	Semi-fruitful	New 9:04 pm
14 Fri 9:18 am	Gemini	Air	Barren	1st
15 Sat	Gemini	Air	Barren	1st
16 Sun 1:46 pm	Cancer	Water	Fruitful	1st
17 Mon	Cancer	Water	Fruitful	1st
18 Tue 5:06 pm	Leo	Fire	Barren	1st
19 Wed	Leo	Fire	Barren	1st
20 Thu 7:58 pm	Virgo	Earth	Barren	2nd 7:43 pm
21 Fri	Virgo	Earth	Barren	2nd
22 Sat 10:50 pm	Libra	Air	Semi-fruitful	2nd
23 Sun	Libra	Air	Semi-fruitful	2nd
24 Mon	Libra	Air	Semi-fruitful	2nd
25 Tue 2:17 am	Scorpio	Water	Fruitful	2nd
26 Wed	Scorpio	Water	Fruitful	2nd
27 Thu 7:15 am	Sagittarius	Fire	Barren	Full 7:07 pm
28 Fri	Sagittarius	Fire	Barren	3rd
29 Sat 2:44 pm	Capricorn	Earth	Semi-fruitful	3rd
30 Sun	Capricorn	Earth	Semi-fruitful	3rd
31 Mon	Capricorn	Earth	Semi-fruitful	3rd

May Aspectarian/Favorable & Unfavorable Days

Date	Sun	Mercury	Venus	Mars	Jupiter	Saturn	Uranus	Neptune	Pluto	Aries	Taurus	Gemini	Cancer	Leo	Virgo	Libra	Scorpio	Sagittarius	Capricorn	Aquarius	Pisces
1					Q					f		U		f	u	f		F		f	u
2		T				Q	Q	X	C	f		U		f	u	f		F		f	u
3	T									u	f		U		f	u	f		F		f
4					X	T	X			u	f		U		f	u	f		F		f
5		Q	T	O						f	u	f		U		f	u	f		F	
6	Q									f	u	f		U		f	u	f		F	
7		X						C	X	f	u	f		U		f	u	f		F	
8	X		Q								f	u	f		U		f	u	f		F
9					C	O	C				f	u	f		U		f	u	f		F
10									Q	F		f	u	f		U		f	u	f	
11			X	T						F		f	u	f		U		f	u	f	
12		C						X	T		F		f	u	f		U		f	u	f
13	C		Q								F		f	u	f		U		f	u	f
14					X	T	X	Q			F		f	u	f		U		f	u	f
15			X							f		F		f	u	f		U		f	u
16		X	C		Q	Q	Q	T	O	f		F		f	u	f		U		f	u
17										u	f		F		f	u	f		U		f
18	X				T	X	T			u	f		F		f	u	f		U		f
19		Q								f	u	f		F		f	u	f		U	
20	Q		X	C				O		f	u	f		F		f	u	f		U	
21		T							T	f	u	f		F		f	u	f		U	
22					O	C	O				f	u	f		F		f	u	f		U
23	T		Q						Q		f	u	f		F		f	u	f		U
24				X						U		f	u	f		F		f	u	f	
25		O	T					T	X	U		f	u	f		F		f	u	f	
26			Q								U		f	u	f		F		f	u	f
27	O					T	X	T	Q		U		f	u	f		F		f	u	f
28										f		U		f	u	f		F		f	u
29				T	Q	Q	Q	X	C	f		U		f	u	f		F		f	u
30		T	O							u	f		U		f	u	f		F		f
31					X	T				u	f		U		f	u	f		F		f

June Moon Table

Date	Sign	Element	Nature	Phase
1 Tue 1:08 am	Aquarius	Air	Barren	3rd
2 Wed	Aquarius	Air	Barren	3rd
3 Thu 1:34 pm	Pisces	Water	Fruitful	3rd
4 Fri	Pisces	Water	Fruitful	4th 6:13 pm
5 Sat	Pisces	Water	Fruitful	4th
6 Sun 1:50 am	Aries	Fire	Barren	4th
7 Mon	Aries	Fire	Barren	4th
8 Tue 11:41 am	Taurus	Earth	Semi-fruitful	4th
9 Wed	Taurus	Earth	Semi-fruitful	4th
10 Thu 6:11 pm	Gemini	Air	Barren	4th
11 Fri	Gemini	Air	Barren	4th
12 Sat 9:50 pm	Cancer	Water	Fruitful	New 7:15 am
13 Sun	Cancer	Water	Fruitful	1st
14 Mon 11:54 pm	Leo	Fire	Barren	1st
15 Tue	Leo	Fire	Barren	1st
16 Wed	Leo	Fire	Barren	1st
17 Thu 1:41 am	Virgo	Earth	Barren	1st
18 Fri	Virgo	Earth	Barren	1st
19 Sat 4:13 am	Libra	Air	Semi-fruitful	2nd 12:30 am
20 Sun	Libra	Air	Semi-fruitful	2nd
21 Mon 8:14 am	Scorpio	Water	Fruitful	2nd
22 Tue	Scorpio	Water	Fruitful	2nd
23 Wed 2:10 pm	Sagittarius	Fire	Barren	2nd
24 Thu	Sagittarius	Fire	Barren	2nd
25 Fri 10:21 pm	Capricorn	Earth	Semi-fruitful	2nd
26 Sat	Capricorn	Earth	Semi-fruitful	Full 7:30 am
27 Sun	Capricorn	Earth	Semi-fruitful	3rd
28 Mon 8:52 am	Aquarius	Air	Barren	3rd
29 Tue	Aquarius	Air	Barren	3rd
30 Wed 9:10 pm	Pisces	Water	Fruitful	3rd

June Aspectarian/Favorable & Unfavorable Days

Date	Sun	Mercury	Venus	Mars	Jupiter	Saturn	Uranus	Neptune	Pluto	Aries	Taurus	Gemini	Cancer	Leo	Virgo	Libra	Scorpio	Sagittarius	Capricorn	Aquarius	Pisces
1							X			f	u	f		U		f	u	f		F	
2	T	Q								f	u	f		U		f	u	f		F	
3			O					C	X	f	u	f		U		f	u	f		F	
4	Q										f	u	f		U		f	u	f		F
5		X	T			O					f	u	f		U		f	u	f		F
6					C		C		Q	F		f	u	f		U		f	u	f	
7	X		Q							F		f	u	f		U		f	u	f	
8			T					X	T	F		f	u	f		U		f	u	f	
9											F		f	u	f		U		f	u	f
10		C	X	Q	X	T	X	Q			F		f	u	f		U		f	u	f
11										f		F		f	u	f		U		f	u
12	C					Q	Q	Q	T	f		F		f	u	f		U		f	u
13				X					O	u	f		F		f	u	f		U		f
14						X				u	f		F		f	u	f		U		f
15		X	C		T		T			f	u	f		F		f	u	f		U	
16	X							O		f	u	f		F		f	u	f		U	
17			C						T	f	u	f		F		f	u	f		U	
18		Q								f	u	f		F		f	u	f		U	
19	Q		X		O	C	O		Q	f	u	f		F		f	u	f		U	
20		T									f	u	f		F		f	u	f		U
21	T			X				T	X		f	u	f		F		f	u	f		U
22			Q							U		f	u	f		F		f	u	f	
23						T	X	T	Q	U		f	u	f		F		f	u	f	
24			T	Q							U		f	u	f		F		f	u	f
25						Q	Q	X			U		f	u	f		F		f	u	f
26	O	O		T	Q				C	f		U		f	u	f		F		f	u
27										f		U		f	u	f		F		f	u
28						X	T	X		f		U		f	u	f		F		f	u
29			O							u	f		U		f	u	f		F		f
30								C		u	f		U		f	u	f		F		f

July Moon Table

Date	Sign	Element	Nature	Phase
1 Thu	Pisces	Water	Fruitful	3rd
2 Fri	Pisces	Water	Fruitful	3rd
3 Sat 9:44 am	Aries	Fire	Barren	3rd
4 Sun	Aries	Fire	Barren	4th 10:35 am
5 Mon 8:29 pm	Taurus	Earth	Semi-fruitful	4th
6 Tue	Taurus	Earth	Semi-fruitful	4th
7 Wed	Taurus	Earth	Semi-fruitful	4th
8 Thu 3:51 am	Gemini	Air	Barren	4th
9 Fri	Gemini	Air	Barren	4th
10 Sat 7:38 am	Cancer	Water	Fruitful	4th
11 Sun	Cancer	Water	Fruitful	New 3:40 pm
12 Mon 8:53 am	Leo	Fire	Barren	1st
13 Tue	Leo	Fire	Barren	1st
14 Wed 9:15 am	Virgo	Earth	Barren	1st
15 Thu	Virgo	Earth	Barren	1st
16 Fri 10:24 am	Libra	Air	Semi-fruitful	1st
17 Sat	Libra	Air	Semi-fruitful	1st
18 Sun 1:42 pm	Scorpio	Water	Fruitful	2nd 6:11 am
19 Mon	Scorpio	Water	Fruitful	2nd
20 Tue 7:48 pm	Sagittarius	Fire	Barren	2nd
21 Wed	Sagittarius	Fire	Barren	2nd
22 Thu	Sagittarius	Fire	Barren	2nd
23 Fri 4:39 am	Capricorn	Earth	Semi-fruitful	2nd
24 Sat	Capricorn	Earth	Semi-fruitful	2nd
25 Sun 3:38 pm	Aquarius	Air	Barren	Full 9:37 pm
26 Mon	Aquarius	Air	Barren	3rd
27 Tue	Aquarius	Air	Barren	3rd
28 Wed 4:00 am	Pisces	Water	Fruitful	3rd
29 Thu	Pisces	Water	Fruitful	3rd
30 Fri 4:42 pm	Aries	Fire	Barren	3rd
31 Sat	Aries	Fire	Barren	3rd

July Aspectarian/Favorable & Unfavorable Days

Date	Sun	Mercury	Venus	Mars	Jupiter	Saturn	Uranus	Neptune	Pluto	Aries	Taurus	Gemini	Cancer	Leo	Virgo	Libra	Scorpio	Sagittarius	Capricorn	Aquarius	Pisces
1	T								X		f	u	f		U		f	u	f		F
2		T		0							f	u	f		U		f	u	f		F
3					C	0	C		Q		f	u	f		U		f	u	f		F
4	Q									F		f	u	f		U		f	u	f	
5		Q	T					X		F		f	u	f		U		f	u	f	
6									T	F		f	u	f		U		f	u	f	
7	X	X	Q	T						F		f	u	f		U		f	u	f	
8					X	T	X	Q		F		f	u	f		U		f	u	f	
9				Q						f		F		f	u	f		U		f	u
10			X		Q	Q	Q	T	0	f		F		f	u	f		U		f	u
11	C			X						u	f	F		f	u	f		U			f
12		C			T	X	T			u	f	F		f	u	f		U			f
13										f	u	f		F		f	u	f		U	
14			C					0	T	f	u	f		F		f	u	f		U	
15	X			C						f	u	f		F		f	u	f		U	
16					0	C	0		Q	f	u	f		F		f	u	f		U	
17		X								U		f	u	f		F		f	u	f	
18	Q							T	X	U		f	u	f		F		f	u	f	
19		Q	X							U		f	u	f		F		f	u	f	
20	T			X		X	T	Q		U		f	u	f		F		f	u	f	
21		Q		T						f		U		f	u	f		F		f	u
22		T		Q						f		U		f	u	f		F		f	u
23					Q	Q	Q	X	C	f		U		f	u	f		F		f	u
24			T							u	f		U		f	u	f		F		f
25	0			T	X	T	X			u	f		U		f	u	f		F		f
26										f	u	f		U		f	u	f		F	
27									C	f	u	f		U		f	u	f		F	
28		0							X	f	u	f		U		f	u	f		F	
29			0							f	u	f		U		f	u	f		F	
30				0	C	0	C		Q	f	u	f		U		f	u	f		F	
31	T									F		f	u	f		U		f	u	f	

August Moon Table

Date	Sign	Element	Nature	Phase
1 Sun	Aries	Fire	Barren	3rd
2 Mon 4:13 am	Taurus	Earth	Semi-fruitful	3rd
3 Tue	Taurus	Earth	Semi-fruitful	4th 12:59 am
4 Wed 12:54 pm	Gemini	Air	Barren	4th
5 Thu	Gemini	Air	Barren	4th
6 Fri 5:50 pm	Cancer	Water	Fruitful	4th
7 Sat	Cancer	Water	Fruitful	4th
8 Sun 7:23 pm	Leo	Fire	Barren	4th
9 Mon	Leo	Fire	Barren	New 11:08 pm
10 Tue 7:01 pm	Virgo	Earth	Barren	1st
11 Wed	Virgo	Earth	Barren	1st
12 Thu 6:43 pm	Libra	Air	Semi-fruitful	1st
13 Fri	Libra	Air	Semi-fruitful	1st
14 Sat 8:26 pm	Scorpio	Water	Fruitful	1st
15 Sun	Scorpio	Water	Fruitful	1st
16 Mon	Scorpio	Water	Fruitful	2nd 2:14 pm
17 Tue 1:34 am	Sagittarius	Fire	Barren	2nd
18 Wed	Sagittarius	Fire	Barren	2nd
19 Thu 10:17 am	Capricorn	Earth	Semi-fruitful	2nd
20 Fri	Capricorn	Earth	Semi-fruitful	2nd
21 Sat 9:37 pm	Aquarius	Air	Barren	2nd
22 Sun	Aquarius	Air	Barren	2nd
23 Mon	Aquarius	Air	Barren	2nd
24 Tue 10:11 am	Pisces	Water	Fruitful	Full 1:05 pm
25 Wed	Pisces	Water	Fruitful	3rd
26 Thu 10:49 pm	Aries	Fire	Barren	3rd
27 Fri	Aries	Fire	Barren	3rd
28 Sat	Aries	Fire	Barren	3rd
29 Sun 10:35 am	Taurus	Earth	Semi-fruitful	3rd
30 Mon	Taurus	Earth	Semi-fruitful	3rd
31 Tue 8:19 pm	Gemini	Air	Barren	3rd

August Aspectarian/Favorable & Unfavorable Days

Date	Sun	Mercury	Venus	Mars	Jupiter	Saturn	Uranus	Neptune	Pluto	Aries	Taurus	Gemini	Cancer	Leo	Virgo	Libra	Scorpio	Sagittarus	Capricorn	Aquarius	Pisces
1								X		F		f	u	f		U		f	u	f	
2		T							T	F		f	u	f		U		f	u	f	
3	Q										F		f	u	f		U		f	u	f
4			T	T	X	T	X	Q			F		f	u	f		U		f	u	f
5	X	Q								f		F		f	u	f		U		f	u
6			Q		Q	Q	Q	T	O	f		F		f	u	f		U		f	u
7		X		Q						u	f		F		f	u	f		U		f
8			X				X	T		u	f		F		f	u	f		U		f
9	C			X	T					f	u	f		F		f	u	f		U	
10								O	T	f	u	f		F		f	u	f		U	
11		C									f	u	f		F		f	u	f		U
12					O	C	O		Q		f	u	f		F		f	u	f		U
13			C	C						U		f	u	f		F		f	u	f	
14	X							T		U		f	u	f		F		f	u	f	
15									X		U		f	u	f		F		f	u	f
16	Q	X						Q			U		f	u	f		F		f	u	f
17			X	X	T	X	T			f		U		f	u	f		F		f	u
18		Q								f		U		f	u	f		F		f	u
19	T				Q	Q	Q	X	C	u	f		U		f	u	f		F		f
20		T	Q	Q						u	f		U		f	u	f		F		f
21							X			u	f		U		f	u	f		F		f
22					X	T				f	u	f		U		f	u	f		F	
23		T	T							f	u	f		U		f	u	f		F	
24	O							C	X	f	u	f		U		f	u	f		F	
25		O									f	u	f		U		f	u	f		F
26						C					f	u	f		U		f	u	f		F
27					C	O			Q	F		f	u	f		U		f	u	f	
28			O	O						F		f	u	f		U		f	u	f	
29	T							X	T	F		f	u	f		U		f	u	f	
30		T									F		f	u	f		U		f	u	f
31				X		X	Q				F		f	u	f		U		f	u	f

September Moon Table

Date	Sign	Element	Nature	Phase
1 Wed	Gemini	Air	Barren	4th 1:22 pm
2 Thu	Gemini	Air	Barren	4th
3 Fri 2:50 am	Cancer	Water	Fruitful	4th
4 Sat	Cancer	Water	Fruitful	4th
5 Sun 5:45 am	Leo	Fire	Barren	4th
6 Mon	Leo	Fire	Barren	4th
7 Tue 5:53 am	Virgo	Earth	Barren	4th
8 Wed	Virgo	Earth	Barren	New 6:30 am
9 Thu 5:01 am	Libra	Air	Semi-fruitful	1st
10 Fri	Libra	Air	Semi-fruitful	1st
11 Sat 5:21 am	Scorpio	Water	Fruitful	1st
12 Sun	Scorpio	Water	Fruitful	1st
13 Mon 8:52 am	Sagittarius	Fire	Barren	1st
14 Tue	Sagittarius	Fire	Barren	1st
15 Wed 4:30 pm	Capricorn	Earth	Semi-fruitful	2nd 1:50 am
16 Thu	Capricorn	Earth	Semi-fruitful	2nd
17 Fri	Capricorn	Earth	Semi-fruitful	2nd
18 Sat 3:35 am	Aquarius	Air	Barren	2nd
19 Sun	Aquarius	Air	Barren	2nd
20 Mon 4:15 pm	Pisces	Water	Fruitful	2nd
21 Tue	Pisces	Water	Fruitful	2nd
22 Wed	Pisces	Water	Fruitful	2nd
23 Thu 4:47 am	Aries	Fire	Barren	Full 5:17 am
24 Fri	Aries	Fire	Barren	3rd
25 Sat 4:17 pm	Taurus	Earth	Semi-fruitful	3rd
26 Sun	Taurus	Earth	Semi-fruitful	3rd
27 Mon	Taurus	Earth	Semi-fruitful	3rd
28 Tue 2:10 am	Gemini	Air	Barren	3rd
29 Wed	Gemini	Air	Barren	3rd
30 Thu 9:46 am	Cancer	Water	Fruitful	4th 11:52 pm

September Aspectarian/Favorable & Unfavorable Days

Date	Sun	Mercury	Venus	Mars	Jupiter	Saturn	Uranus	Neptune	Pluto	Aries	Taurus	Gemini	Cancer	Leo	Virgo	Libra	Scorpio	Sagittarus	Capricorn	Aquarius	Pisces
1	Q	Q				T				f		F		f	u	f		U		f	u
2		T	T					T		f		F		f	u	f		U		f	u
3	X	X			Q	Q	Q		O	u	f		F		f	u	f		U		f
4			Q							u	f		F		f	u	f		U		f
5		Q			T	X	T			u	f		F		f	u	f		U		f
6				X						f	u	f		F		f	u	f		U	
7		X						O	T	f	u	f		F		f	u	f		U	
8	C										f	u	f		F		f	u	f		U
9					O	C	O		Q		f	u	f		F		f	u	f		U
10								T		U		f	u	f		F		f	u	f	
11		X	C	C					X	U		f	u	f		F		f	u	f	
12	X										U		f	u	f		F		f	u	f
13		Q			T	X	T	Q			U		f	u	f		F		f	u	f
14										f		U		f	u	f		F		f	u
15	Q			X	Q		Q	X	C	f		U		f	u	f		F		f	u
16		T	X			Q				u	f		U		f	u	f		F		f
17	T									u	f		U		f	u	f		F		f
18			Q	Q	X	T	X			u	f		U		f	u	f		F		f
19										f	u	f		U		f	u	f		F	
20								C	X	f	u	f		U		f	u	f		F	
21		O	T	T							f	u	f		U		f	u	f		F
22											f	u	f		U		f	u	f		F
23	O				C	O	C		Q		f	u	f		U		f	u	f		F
24										F		f	u	f		U		f	u	f	
25								X	T	F		f	u	f		U		f	u	f	
26			O	O							F		f	u	f		U		f	u	f
27		T			X		X	Q			F		f	u	f		U		f	u	f
28	T					T				f		F		f	u	f		U		f	u
29		Q								f		F		f	u	f		U		f	u
30	Q				Q	Q	Q	T	O	f		F		f	u	f		U		f	u

October Moon Table

Date	Sign	Element	Nature	Phase
1 Fri	Cancer	Water	Fruitful	4th
2 Sat 2:21 pm	Leo	Fire	Barren	4th
3 Sun	Leo	Fire	Barren	4th
4 Mon 4:00 pm	Virgo	Earth	Barren	4th
5 Tue	Virgo	Earth	Barren	4th
6 Wed 3:52 pm	Libra	Air	Semi-fruitful	4th
7 Thu	Libra	Air	Semi-fruitful	New 2:45 pm
8 Fri 3:52 pm	Scorpio	Water	Fruitful	1st
9 Sat	Scorpio	Water	Fruitful	1st
10 Sun 6:09 pm	Sagittarius	Fire	Barren	1st
11 Mon	Sagittarius	Fire	Barren	1st
12 Tue	Sagittarius	Fire	Barren	1st
13 Wed 12:17 am	Capricorn	Earth	Semi-fruitful	1st
14 Thu	Capricorn	Earth	Semi-fruitful	2nd 5:27 pm
15 Fri 10:24 am	Aquarius	Air	Barren	2nd
16 Sat	Aquarius	Air	Barren	2nd
17 Sun 10:52 pm	Pisces	Water	Fruitful	2nd
18 Mon	Pisces	Water	Fruitful	2nd
19 Tue	Pisces	Water	Fruitful	2nd
20 Wed 11:23 am	Aries	Fire	Barren	2nd
21 Thu	Aries	Fire	Barren	2nd
22 Fri 10:30 pm	Taurus	Earth	Semi-fruitful	Full 9:37 pm
23 Sat	Taurus	Earth	Semi-fruitful	3rd
24 Sun	Taurus	Earth	Semi-fruitful	3rd
25 Mon 7:47 am	Gemini	Air	Barren	3rd
26 Tue	Gemini	Air	Barren	3rd
27 Wed 3:14 pm	Cancer	Water	Fruitful	3rd
28 Thu	Cancer	Water	Fruitful	3rd
29 Fri 8:39 pm	Leo	Fire	Barren	3rd
30 Sat	Leo	Fire	Barren	4th 8:46 am
31 Sun 11:51 pm	Virgo	Earth	Barren	4th

October Aspectarian/Favorable & Unfavorable Days

Date	Sun	Mercury	Venus	Mars	Jupiter	Saturn	Uranus	Neptune	Pluto	Aries	Taurus	Gemini	Cancer	Leo	Virgo	Libra	Scorpio	Sagittarius	Capricorn	Aquarius	Pisces
1			T	T						u	f		F		f	u	f		U		f
2		X			T			T		u	f		F		f	u	f		U		f
3	X		Q	Q		X				f	u	f		F		f	u	f		U	
4								O	T	f	u	f		F		f	u	f		U	
5			X	X							f	u	f		F		f	u	f		U
6					O		O		Q		f	u	f		F		f	u	f		U
7	C	C				C				U		f	u	f		F		f	u	f	
8								T	X	U		f	u	f		F		f	u	f	
9			C	C							U		f	u	f		F		f	u	f
10					T		T	Q			U		f	u	f		F		f	u	f
11		X				X				f		U		f	u	f		F		f	u
12	X				Q		Q	X		f		U		f	u	f		F		f	u
13						Q			C	u	f		U		f	u	f		F		f
14	Q	Q	X	X						u	f		U		f	u	f		F		f
15					X		X			u	f		U		f	u	f		F		f
16		Q				T				f	u	f		U		f	u	f		F	
17	T	T		Q				C		f	u	f		U		f	u	f		F	
18			T						X		f	u	f		U		f	u	f		F
19			T								f	u	f		U		f	u	f		F
20				C			C	Q			f	u	f		U		f	u	f		F
21						O				F		f	u	f		U		f	u	f	
22	O							X		F		f	u	f		U		f	u	f	
23		O	O						T		F		f	u	f		U		f	u	f
24				X							F		f	u	f		U		f	u	f
25			O				X	Q			F		f	u	f		U		f	u	f
26						T				f		F		f	u	f		U		f	u
27	T				Q		Q	T	O	f		F		f	u	f		U		f	u
28		T	T			Q				u	f		F		f	u	f		U		f
29				T	T		T			u	f		F		f	u	f		U		f
30	Q		Q			X				f	u	f		F		f	u	f		U	
31		Q						O		f	u	f		F		f	u	f		U	

November Moon Table

Date	Sign	Element	Nature	Phase
1 Mon	Virgo	Earth	Barren	4th
2 Tue	Virgo	Earth	Barren	4th
3 Wed 1:19 am	Libra	Air	Semi-fruitful	4th
4 Thu	Libra	Air	Semi-fruitful	4th
5 Fri 2:16 am	Scorpio	Water	Fruitful	4th
6 Sat	Scorpio	Water	Fruitful	New 12:52 am
7 Sun 3:28 am	Sagittarius	Fire	Barren	1st
8 Mon	Sagittarius	Fire	Barren	1st
9 Tue 8:37 am	Capricorn	Earth	Semi-fruitful	1st
10 Wed	Capricorn	Earth	Semi-fruitful	1st
11 Thu 5:32 pm	Aquarius	Air	Barren	1st
12 Fri	Aquarius	Air	Barren	1st
13 Sat	Aquarius	Air	Barren	2nd 11:39 am
14 Sun 5:24 am	Pisces	Water	Fruitful	2nd
15 Mon	Pisces	Water	Fruitful	2nd
16 Tue 5:59 pm	Aries	Fire	Barren	2nd
17 Wed	Aries	Fire	Barren	2nd
18 Thu	Aries	Fire	Barren	2nd
19 Fri 5:04 am	Taurus	Earth	Semi-fruitful	2nd
20 Sat	Taurus	Earth	Semi-fruitful	2nd
21 Sun 1:46 pm	Gemini	Air	Barren	Full 12:27 pm
22 Mon	Gemini	Air	Barren	3rd
23 Tue 8:14 pm	Cancer	Water	Fruitful	3rd
24 Wed	Cancer	Water	Fruitful	3rd
25 Thu	Cancer	Water	Fruitful	3rd
26 Fri 1:01 am	Leo	Fire	Barren	3rd
27 Sat	Leo	Fire	Barren	3rd
28 Sun 4:34 am	Virgo	Earth	Barren	4th 3:36 pm
29 Mon	Virgo	Earth	Barren	4th
30 Tue 7:15 am	Libra	Air	Semi-fruitful	4th

November Aspectarian/Favorable & Unfavorable Days

Date	Sun	Mercury	Venus	Mars	Jupiter	Saturn	Uranus	Neptune	Pluto	Aries	Taurus	Gemini	Cancer	Leo	Virgo	Libra	Scorpio	Sagittarus	Capricorn	Aquarius	Pisces
1	X		X	Q					T		f	u	f		F		f	u	f		U
2		X			O		O				f	u	f		F		f	u	f		U
3				X		C			Q	U		f	u	f		F		f	u	f	
4								T		U		f	u	f		F		f	u	f	
5			C						X		U		f	u	f		F		f	u	f
6	C	C			T		T	Q			U		f	u	f		F		f	u	f
7			C								U		f	u	f		F		f	u	f
8					Q	X				f		U		f	u	f		F		f	u
9			X				Q	X	C	f		U		f	u	f		F		f	u
10	X					Q				u	f		U		f	u	f		F		f
11			Q	X			X			u	f		U		f	u	f		F		f
12		X		X		T				f	u	f		U		f	u	f		F	
13	Q							C		f	u	f		U		f	u	f		F	
14			T						X	f	u	f		U		f	u	f		F	
15		Q	Q								f	u	f		U		f	u	f		F
16	T					C	C				f	u	f		U		f	u	f		F
17		T				O			Q	F		f	u	f		U		f	u	f	
18			T				X			F		f	u	f		U		f	u	f	
19		O							T	F		f	u	f		U		f	u	f	
20											F		f	u	f		U		f	u	f
21	O			X		X	Q				F		f	u	f		U		f	u	f
22					T					f		F		f	u	f		U		f	u
23		O	T	O	Q		Q	T		f		F		f	u	f		U		f	u
24					Q				O	u	f		F		f	u	f		U		f
25		Q		T		T				u	f		F		f	u	f		U		f
26	T									f	u	f		F		f	u	f		U	
27		T		T	X			O		f	u	f		F		f	u	f		U	
28	Q		X						T	f	u	f		F		f	u	f		U	
29				Q	O						f	u	f		F		f	u	f		U
30	X	Q					O		Q		f	u	f		F		f	u	f		U

December Moon Table

Date	Sign	Element	Nature	Phase
1 Wed	Libra	Air	Semi-fruitful	4th
2 Thu 9:44 am	Scorpio	Water	Fruitful	4th
3 Fri	Scorpio	Water	Fruitful	4th
4 Sat 12:59 pm	Sagittarius	Fire	Barren	4th
5 Sun	Sagittarius	Fire	Barren	New 12:36 pm
6 Mon 6:16 pm	Capricorn	Earth	Semi-fruitful	1st
7 Tue	Capricorn	Earth	Semi-fruitful	1st
8 Wed	Capricorn	Earth	Semi-fruitful	1st
9 Thu 2:30 am	Aquarius	Air	Barren	1st
10 Fri	Aquarius	Air	Barren	1st
11 Sat 1:41 pm	Pisces	Water	Fruitful	1st
12 Sun	Pisces	Water	Fruitful	1st
13 Mon	Pisces	Water	Fruitful	2nd 8:59 am
14 Tue 2:15 am	Aries	Fire	Barren	2nd
15 Wed	Aries	Fire	Barren	2nd
16 Thu 1:49 pm	Taurus	Earth	Semi-fruitful	2nd
17 Fri	Taurus	Earth	Semi-fruitful	2nd
18 Sat 10:37 pm	Gemini	Air	Barren	2nd
19 Sun	Gemini	Air	Barren	2nd
20 Mon	Gemini	Air	Barren	2nd
21 Tue 4:22 am	Cancer	Water	Fruitful	Full 3:13 am
22 Wed	Cancer	Water	Fruitful	3rd
23 Thu 7:51 am	Leo	Fire	Barren	3rd
24 Fri	Leo	Fire	Barren	3rd
25 Sat 10:14 am	Virgo	Earth	Barren	3rd
26 Sun	Virgo	Earth	Barren	3rd
27 Mon 12:38 pm	Libra	Air	Semi-fruitful	4th 11:18 pm
28 Tue	Libra	Air	Semi-fruitful	4th
29 Wed 3:49 pm	Scorpio	Water	Fruitful	4th
30 Thu	Scorpio	Water	Fruitful	4th
31 Fri 8:21 pm	Sagittarius	Fire	Barren	4th

December Aspectarian/Favorable & Unfavorable Days

Date	Sun	Mercury	Venus	Mars	Jupiter	Saturn	Uranus	Neptune	Pluto	Aries	Taurus	Gemini	Cancer	Leo	Virgo	Libra	Scorpio	Sagittarius	Capricorn	Aquarius	Pisces
1						C				U		f	u	f		F		f	u	f	
2		X	C	X				T	X	U		f	u	f		F		f	u	f	
3											U		f	u	f		F		f	u	f
4					T		T	Q			U		f	u	f		F		f	u	f
5	C					X				f		U		f	u	f		F		f	u
6				C	Q		Q	X		f		U		f	u	f		F		f	u
7		C	X			Q			C	u	f		U		f	u	f		F		f
8						X		X		u	f		U		f	u	f		F		f
9			Q							f	u	f		U		f	u	f		F	
10	X					T				f	u	f		U		f	u	f		F	
11				X				C	X	f	u	f		U		f	u	f		F	
12		X	T								f	u	f		U		f	u	f		F
13	Q				C		C				f	u	f		U		f	u	f		F
14		Q		Q					Q	F		f	u	f		U		f	u	f	
15						O				F		f	u	f		U		f	u	f	
16	T	T						X	T	F		f	u	f		U		f	u	f	
17			O	T							F		f	u	f		U		f	u	f
18						X		X	Q		F		f	u	f		U		f	u	f
19										f		F		f	u	f		U		f	u
20		O			Q	T	Q	T		f		F		f	u	f		U		f	u
21	O			O					O	f		F		f	u	f		U		f	u
22			T		Q					u	f		F		f	u	f		U		f
23					T		T			u	f		F		f	u	f		U		f
24		T	Q			X				f	u	f		F		f	u	f		U	
25	T						O	T		f	u	f		F		f	u	f		U	
26		Q	X	T							f	u	f		F		f	u	f		U
27	Q				O		O		Q		f	u	f		F		f	u	f		U
28		X		Q		C				U		f	u	f		F		f	u	f	
29								T		U		f	u	f		F		f	u	f	
30	X			X					X		U		f	u	f		F		f	u	f
31			C		T		T	Q			U		f	u	f		F		f	u	f

2010 Retrograde Planets

Planet	Begin	Eastern	Pacific	End	Eastern	Pacific
Mars	12/20/09	8:26 am	**5:26 am**	03/10/10	12:09 pm	**9:09 am**
Mercury	12/26/09	9:38 am	**6:38 am**	01/15/10	11:52 am	**8:52 am**
Saturn	01/13/10	10:56 am	**7:56 am**	05/30/10	2:09 pm	**11:09 am**
Pluto	04/06/10	10:34 pm	**7:34 pm**	09/13/10		**9:37 pm**
				09/14/10	12:37 am	
Mercury	04/17/10		**9:06 pm**	05/11/10	6:27 pm	**3:27 pm**
	04/18/10	12:06 am				
Neptune	05/31/10	2:48 pm	**11:48 pm**	11/06/10		**11:04 pm**
				11/07/10	1:04 am	
Uranus	07/05/10	12:50 pm	**9:50 am**	12/05/10	8:50 pm	**5:50 pm**
Jupiter	07/23/10	8:03 am	**5:03 am**	11/18/10	11:53 am	**8:53 am**
Mercury	08/20/10	3:59 pm	**12:59 pm**	09/12/10	7:09 pm	**4:09 pm**
Venus	10/08/10	3:05 am	**12:05 am**	11/18/10	4:18 pm	**1:18 pm**
Mercury	12/10/10	7:04 am	**4:04 am**	12/29/10		**11:21 pm**
				12/30/10	2:21 am	

Eastern Time in plain type, **Pacific Time in bold type**

	09 Dec	10 Jan	Feb	Mar	Apr	May	Jun	Jul	Aug	Sep	Oct	Nov	Dec	11 Jan
☿		▨			▨				▨				▨	
♀											▨			
♂		▨▨▨												
♃									▨▨▨					
♄			▨▨▨▨											
♅						▨▨▨▨▨								
♆					▨▨▨▨▨									
♇				▨▨▨▨										

Egg-setting Dates

To Have Eggs by this Date	Sign	Qtr.	Date to Set Eggs
Jan 18, 1:17 am–Jan 20, 1:36 pm	Pisces	1st	Dec 28, 2009
Jan 22, 11:39 pm–Jan 25, 6:11 am	Taurus	1st	Jan 1
Jan 27, 9:01 am–Jan 29, 9:10 am	Cancer	2nd	Jan 6
Feb 14, 7:23 am–Feb 16, 7:30 pm	Pisces	1st	Jan 24
Feb 19, 5:55 am–Feb 21, 1:47 pm	Taurus	1st	Jan 29
Feb 23, 6:29 pm–Feb 25, 8:08 pm	Cancer	2nd	Feb 2
Mar 15, 5:01 pm–Mar 16, 2:32 am	Pisces	1st	Feb 22
Mar 18, 12:29 pm–Mar 20, 8:28 pm	Taurus	1st	Feb 25
Mar 23, 2:16 am–Mar 25, 5:39 am	Cancer	1st	Mar 2
Mar 29, 7:21 am–Mar 29, 10:25 pm	Libra	2nd	Mar 8
Apr 14, 6:55 pm–Apr 17, 2:08 am	Taurus	1st	Mar 24
Apr 19, 7:39 am–Apr 21, 11:42 am	Cancer	1st	Mar 29
Apr 25, 4:16 pm–Apr 27, 6:28 pm	Libra	2nd	Apr 4
May 13, 9:04 pm–May 14, 9:18 am	Taurus	1st	Apr 22
May 16, 1:46 pm–May 18, 5:06 pm	Cancer	1st	Apr 25
May 22, 10:50 pm–May 25, 2:17 am	Libra	2nd	May 1
Jun 12, 9:50 pm–Jun 14, 11:54 pm	Cancer	1st	May 22
Jun 19, 4:13 am–Jun 21, 8:14 am	Libra	2nd	May 29
Jul 11, 3:40 pm–Jul 12, 8:53 am	Cancer	1st	Jun 20
Jul 16, 10:24 am–Jul 18, 1:42 pm	Libra	1st	Jun 25
Aug 12, 6:43 pm–Aug 14, 8:26 pm	Libra	1st	Jul 22
Aug 24, 10:11 am–Aug 24, 1:05 pm	Pisces	2nd	Aug 3
Sep 9, 5:01 am–Sep 11, 5:21 am	Libra	1st	Aug 19
Sep 20, 4:15 pm–Sep 23, 4:47 am	Pisces	2nd	Aug 30
Oct 7, 2:45 pm–Oct 8, 3:52 pm	Libra	1st	Sep 16
Oct 17, 10:52 pm–Oct 20, 11:23 am	Pisces	2nd	Sep 26
Nov 14, 5:24 am–Nov 16, 5:59 pm	Pisces	2nd	Oct 24
Nov 19, 5:04 am–Nov 21, 12:27 pm	Taurus	2nd	Oct 29
Dec 11, 1:41 pm–Dec 14, 2:15 am	Pisces	1st	Nov 20
De. 16, 1:49 pm–Dec 18, 10:37 pm	Taurus	2nd	Nov 25

Dates to Hunt and Fish

Date	Quarter	Sign
Jan 8, 5:00 am–Jan 10, 1:10 pm	4th	Scorpio
Jan 18, 1:17 am–Jan 20, 1:36 pm	1st	Pisces
Jan 27, 9:01 am–Jan 29, 9:10 am	2nd	Cancer
Feb 4, 11:56 am–Feb 6, 7:04 pm	3rd	Scorpio
Feb 14, 7:23 am–Feb 16, 7:30 pm	1st	Pisces
Feb 23, 6:29 am–Feb 25, 8:08 pm	2nd	Cancer
Mar 3, 9:11 am–Mar 6, 2:36 am	3rd	Scorpio
Mar 6, 2:36 am–Mar 8, 12:13 pm	3rd	Sagittarius
Mar 13, 1:44 pm–Mar 16, 2:32 am	4th	Pisces
Mar 23, 2:16 am–Mar 25, 5:39 am	1st	Cancer
Mar 31, 8:41 am–Apr 2, 12:53 pm	3rd	Scorpio
Apr 2, 12:53 pm–Apr 4, 9:07 pm	3rd	Sagittarius
Apr 9, 9:48 pm–Apr 12, 9:31 am	4th	Pisces
Apr 19, 7:39 am–Apr 21, 11:42 am	1st	Cancer
Apr 27, 6:28 pm–Apr 29, 10:36 pm	2nd	Scorpio
Apr 29, 10:36 pm–May 2, 6:00 am	3rd	Sagittarius
May 7, 5:34 am–May 9, 5:29 pm	4th	Pisces
May 16, 1:46 pm–May 18, 5:06 pm	1st	Cancer
May 25, 2:17 am–May 27, 7:15 am	2nd	Scorpio
May 27, 7:15 am–May 29, 2:44 pm	2nd	Sagittarius
Jun 3, 1:34 pm–Jun 6, 1:50 am	3rd	Pisces
Jun 12, 9:50 pm–Jun 14, 11:54 pm	1st	Cancer
Jun 21, 8:14 am–Jun 23, 2:10 pm	2nd	Scorpio
Jun 23, 2:10 pm–Jun 25, 10:21 pm	2nd	Sagittarius
Jun 30, 9:10 am–Jul 3, 9:44 am	3rd	Pisces
Jul 3, 9:44 am–Jul 5, 8:29 pm	3rd	Aries
Jul 10, 7:38 am–Jul 12, 8:53 am	4th	Cancer
Jul 18, 1:42 pm–Jul 20, 7:48 pm	2nd	Scorpio
Jul 20, 7:48 pm–Jul 23, 4:39 am	2nd	Sagittarius
Jul 28, 4:00 am–Jul 30, 4:42 pm	3rd	Pisces
Jul 30, 4:42 pm–Aug 2, 4:13 am	3rd	Aries
Aug 6, 5:50 am–Aug 8, 7:23 pm	4th	Cancer
Aug 14, 8:26 pm–Aug 17, 1:34 am	1st	Scorpio
Aug 17, 1:34 am–Aug 19, 10:17 am	2nd	Sagittarius
Aug 24, 10:11 am–Aug 26, 10:49 pm	2nd	Pisces
Aug 26, 10:49 pm–Aug 29, 10:35 am	3rd	Aries
Sep 3, 2:50 am–Sep 5, 5:45 am	4th	Cancer
Sep 11, 5:21 am–Sep 13, 8:52 am	1st	Scorpio
Sep 20, 4:15 pm–Sep 23, 4:47 am	2nd	Pisces
Sep 23, 4:47 am–Sep 25, 4:17 pm	2nd	Aries
Sep 30, 9:46 am–Oct 2, 2:21 pm	3rd	Cancer
Oct 8, 3:52 pm–Oct 10, 6:09 pm	1st	Scorpio
Oct 17, 10:52 pm–Oct 20, 11:23 am	2nd	Pisces
Oct 20, 11:23 am–Oct 22, 10:30 pm	2nd	Aries
Oct 27, 3:14 pm–Oct 29, 8:39 pm	3rd	Cancer
Nov 5, 2:16 am–Nov 7, 3:28 am	4th	Scorpio
Nov 14, 5:24 am–Nov 16, 5:59 pm	2nd	Pisces
Nov 16, 5:59 pm–Nov 19, 5:04 am	2nd	Aries
Nov 23, 8:14 pm–Nov 26, 1:01 am	3rd	Cancer
Dec 2, 9:44 am–Dec 4, 12:59 pm	4th	Scorpio
Dec 11, 1:41 pm–Dec 14, 2:15 am	1st	Pisces
Dec 14, 2:15 am–Dec 16, 1:49 pm	2nd	Aries
Dec 21, 4:22 am–Dec 23, 7:51 pm	3rd	Cancer
Dec 29, 3:49 pm–Dec 31, 8:21 pm	4th	Scorpio

Dates to Destroy Weeds and Pests

From		To		Sign	Qtr.
Jan 1	9:41 pm	Jan 3	9:52 pm	Leo	3rd
Jan 3	9:52 pm	Jan 5	11:58 pm	Virgo	3rd
Jan 10	1:10 pm	Jan 12	11:54 pm	Sagittarius	4th
Jan 30	1:18 am	Jan 31	8:23 am	Leo	3rd
Jan 31	8:23 am	Feb 2	8:42 am	Virgo	3rd
Feb 6	7:04 pm	Feb 9	5:43 am	Sagittarius	4th
Feb 11	6:24 pm	Feb 13	9:51 pm	Aquarius	4th
Feb 28	11:38 am	Mar 1	7:31 pm	Virgo	3rd
Mar 6	2:36 am	Mar 7	10:42 am	Sagittarius	3rd
Mar 7	10:42 am	Mar 8	12:13 pm	Sagittarius	4th
Mar 11	12:42 am	Mar 13	1:44 pm	Aquarius	4th
Apr 2	12:53 pm	Apr 4	9:07 pm	Sagittarius	3rd
Apr 7	8:51 am	Apr 9	9:48 pm	Aquarius	4th
Apr 12	9:31 am	Apr 14	8:29 am	Aries	4th
Apr 29	10:36 pm	May 2	6:00 am	Sagittarius	3rd
May 4	4:52 pm	May 6	12:15 am	Aquarius	3rd
May 6	12:15 am	May 7	5:34 am	Aquarius	4th
May 9	5:29 pm	May 12	2:48 am	Aries	4th
May 27	7:07 pm	May 29	2:44 am	Sagittarius	3rd
Jun 1	1:08 am	Jun 3	1:34 pm	Aquarius	3rd
Jun 6	1:50 am	Jun 8	11:41 am	Aries	4th
Jun 10	6:11 pm	Jun 12	7:15 am	Gemini	4th
Jun 28	8:52 am	Jun 30	9:10 pm	Aquarius	3rd
Jul 3	9:44 am	Jul 4	10:35 am	Aries	3rd
Jul 4	10:35 am	Jul 5	8:29 pm	Aries	4th
Jul 8	3:51 am	Jul 10	7:38 am	Gemini	4th
Jul 25	9:37 pm	Jul 28	4:00 am	Aquarius	3rd
Jul 30	4:42 pm	Aug 2	4:13 am	Aries	3rd
Aug 4	12:54 pm	Aug 6	5:50 am	Gemini	4th
Aug 8	7:23 pm	Aug 9	11:08 pm	Leo	4th
Aug 26	10:49 pm	Aug 29	10:35 am	Aries	3rd
Aug 31	8:19 pm	Sep 1	1:22 pm	Gemini	3rd
Sep 1	1:22 pm	Sep 3	2:50 am	Gemini	4th
Sep 5	5:45 am	Sep 7	5:53 am	Leo	4th
Sep 7	5:53 am	Sep 8	6:30 am	Virgo	4th
Sep 23	5:17 am	Sep 25	4:17 pm	Aries	3rd
Sep 28	2:10 am	Sep 30	9:46 am	Gemini	3rd
Oct 2	2:21 pm	Oct 4	4:00 pm	Leo	4th
Oct 4	4:00 pm	Oct 6	3:52 pm	Virgo	4th
Oct 22	9:37 pm	Oct 22	10:30 pm	Aries	3rd
Oct 25	7:47 am	Oct 27	3:14 pm	Gemini	3rd
Oct 29	8:39 pm	Oct 30	8:46 am	Leo	3rd
Oct 30	8:46 am	Oct 31	11:51 pm	Leo	4th
Oct 31	11:51 pm	Nov 3	1:19 am	Virgo	4th
Nov 21	1:46 pm	Nov 23	8:14 pm	Gemini	3rd
Nov 26	1:01 am	Nov 28	4:34 am	Leo	3rd
Nov 28	4:34 am	Nov 28	3:36 pm	Virgo	3rd
Nov 28	3:36 pm	Nov 30	7:15 am	Virgo	4th
Dec 4	12:59 pm	Dec 5	12:36 pm	Sagittarius	4th
Dec 21	3:13 am	Dec 21	4:22 am	Gemini	3rd
Dec 23	7:51 am	Dec 25	10:14 am	Leo	3rd
Dec 25	10:14 am	Dec 27	12:38 pm	Virgo	3rd

Time Zone Map

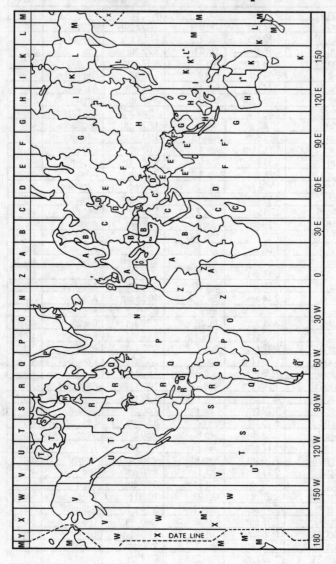

Time Zone Conversions

(R) EST—Used in book
(S) CST—Subtract 1 hour
(T) MST—Subtract 2 hours
(U) PST—Subtract 3 hours
(V) Subtract 4 hours
(V*) Subtract 4½ hours
(U*) Subtract 3½ hours
(W) Subtract 5 hours
(X) Subtract 6 hours
(Y) Subtract 7 hours
(Q) Add 1 hour
(P) Add 2 hours
(P*) Add 2½ hours
(O) Add 3 hours
(N) Add 4 hours
(Z) Add 5 hours
(A) Add 6 hours
(B) Add 7 hours
(C) Add 8 hours
(C*) Add 8½ hours

(D) Add 9 hours
(D*) Add 9½ hours
(E) Add 10 hours
(E*) Add 10½ hours
(F) Add 11 hours
(F*) Add 11½ hours
(G) Add 12 hours
(H) Add 13 hours
(I) Add 14 hours
(I*) Add 14½ hours
(K) Add 15 hours
(K*) Add 15½ hours
(L) Add 16 hours
(L*) Add 16½ hours
(M) Add 17 hours
(M*) Add 18 hours
(P*) Add 2½ hours

Important!

All times given in the *Moon Sign Book* are set in Eastern Time. The conversions shown here are for standard times only. Use the time zone conversions map and table to calculate the difference in your time zone. You must make the adjustment for your time zone and adjust for Daylight Saving Time where applicable.

Weather, Economic, & Lunar Forecasts

Forecasting the Weather

By Kris Brandt Riske

Astrometeorology—astrological weather forecasting—reveals seasonal and weekly weather trends based on the cardinal ingresses (Summer and Winter Solstices, and Spring and Autumn Equinoxes) and the four monthly lunar phases. The planetary alignments and the longitudes and latitudes they influence have the strongest effect, but the zodiacal signs are also involved in creating weather conditions.

The components of a thunderstorm, for example, are heat, wind, and electricity. A Mars-Jupiter configuration generates the necessary heat and Mercury adds wind and electricity. A severe thunderstorm, and those that produce tornados, usually involve Mercury, Mars, Uranus, or Neptune. The zodiacal signs add their energy to the planetary mix to increase or decrease the chance of weather phenomena and their severity.

In general, the fire signs (Aries, Leo, Sagittarius) indicate heat and dryness, both of which peak when Mars, the planet with a similar nature, is in these signs. Water signs (Cancer, Scorpio, Pisces) are conducive to precipitation, and air signs (Gemini, Libra, Aquarius) to cool temperatures and wind. Earth signs (Taurus, Virgo, Capricorn) vary from wet to dry, heat to cold. The signs and their prevailing weather conditions are listed here:

Aries: Heat, dry, wind
Taurus: Moderate temperatures, precipitation
Gemini: Cool temperatures, wind, dry
Cancer: Cold, steady precipitation
Leo: Heat, dry, lightning
Virgo: Cold, dry, windy
Libra: Cool, windy, fair
Scorpio: Extreme temperatures, abundant precipitation
Sagittarius: Warm, fair, moderate wind
Capricorn: Cold, wet, damp
Aquarius: Cold, dry, high pressure, lightning
Pisces: Wet, cool, low pressure

Take note of the Moon's sign at each lunar phase. It reveals the prevailing weather conditions for the next six to seven days. The same is true of Mercury and Venus. These two influential weather planets transit the entire zodiac each year, unless retrograde patterns add their influence.

Planetary Influences

People relied on astrology to forecast weather for thousands of years. They were able to predict drought, floods, and temperature variations through interpreting planetary alignments. In recent years there has been a renewed interest in astrometeorology. A weather forecast can be composed for any date—tomorrow, next week, or a thousand years in the future. According to astrometeorology, each planet governs certain weather phenomena. When certain planets are aligned with other planets, weather—precipitation, cloudy or clear skies, tornados, hurricanes, and other conditions—are generated.

Sun and Moon

The Sun governs the constitution of the weather and, like the Moon, it serves as a trigger for other planetary configurations that result in weather events. When the Sun is prominent in a cardinal ingress or lunar phase chart, the area is often warm and sunny. The Moon can bring or withhold moisture, depending upon its sign placement.

Mercury

Mercury is also a triggering planet, but its main influence is wind direction and velocity. In its stationary periods, Mercury reflects high winds, and its influence is always prominent in major weather events, such as hurricanes and tornados, when it tends to lower the temperature.

Venus

Venus governs moisture, clouds, and humidity. It brings warming trends that produce sunny, pleasant weather if in positive aspect to other planets. In some signs—Libra, Virgo, Gemini, Sagittarius—Venus is drier. It is at its wettest when placed in Cancer, Scorpio, Pisces, or Taurus.

Mars

Mars is associated with heat, drought, and wind, and can raise the temperature to record-setting levels when in a fire sign (Aries, Leo, Sagittarius). Mars ios also the planet that provides the spark that generates thunderstorms and is prominent in tornado and hurricane configurations.

Jupiter

Jupiter, a fair-weather planet, tends toward higher temperatures when in Aries, Leo, or Sagittarius. It is associated with high-pressure systems and is a contributing factor at times to dryness. Storms are often amplified by Jupiter.

Saturn

Saturn is associated with low-pressure systems, cloudy to overcast skies, and excessive precipitation. Temperatures drop when Saturn is involved. Major winter storms always have a strong Saturn influence, as do storms that produce a slow, steady downpour for hours or days.

Uranus

Like Jupiter, Uranus indicates high-pressure systems. It reflects descending cold air and, when prominent, is responsible for a jet stream that extends far south. Uranus can bring drought in winter, and it is involved in thunderstorms, tornados, and hurricanes.

Neptune

Neptune is the wettest planet. It signals low-pressure systems and is dominant when hurricanes are in the forecast. When Neptune is strongly placed, flood danger is high. It's often associated with winter thaws. Temperatures, humidity, and cloudiness increase where Neptune influences weather.

Pluto

Pluto is associated with weather extremes, as well as unseasonably warm temperatures and drought. It reflects the high winds involved in major hurricanes, storms, and tornados.

Meteorology and Astrometeorology: The Dynamic Duo of Weather Forecasting

By Kris Brandt Riske

Meteorology is the science of weather forecasting. Astrometeorology is the astrology of weather forecasting. Each has its strengths and weaknesses, and there are similarities as well as differences between the two methods. Meteorologists use physics and thermodynamics (changes in air pressure, volume, and temperature) to forecast weather. Astrometeorlogists use planets, signs, solar ingresses, and lunar phases to do so. Although astrometeorology is obviously symbolic in nature (for example, Mercury represents wind), operational meteorology is somewhat the same in that a high pressure, for example, generally indicates fair weather. Science explains why meteorology "works," but there is, as of yet, no known reason why astrometeorology works.

History of Meteorology and Astrometeorology

Meteorology has its roots in the fourth century BCE, when Aristotle explained his theories on water vapor and weather forecasting in *Meteorologica*. But it wasn't until the seventeenth and eighteenth centuries that meteorology became a discipline, with the invention of thermometers and barometers. By the nineteenth century, meteorological observations were being recorded in some locations, and in the 1920s, theories for fronts, air masses, and mid-latitude cyclones were developed. Radar, common today, was first used in World War II, and the 1950s saw the first atmospheric computer models, followed by the first meteorological satellites in the 1960s. It's only been in the past twenty or so years, however, that the multitude of complex, computer-generated tools—used daily by meteorologists—have been developed and refined in order to produce more accurate short-range forecasts. In this context, meteorology is still a very "young" science, although one that is evolving rapidly.

Astrometeorology—which, like astrology, is based on the science of astronomy—was probably first employed 2,500 years ago, when the ancients correlated the movement of the planets to events on Earth. Ptolemy, an astrologer-astronomer of the second century, used astrology to predict the weather. But it wasn't until the seventeenth century that the rules for astrometeorology were recorded in one volume—Dr. J. Goad's *Astro-Meteorologica* in 1686. Since then, the popularity of astrometeorology has waxed and waned, but it has been a staple of *Llewellyn's Moon Sign Book* since 1905.

Although some might say the recent advances in meteorology make astrometeorology obsolete, the opposite is true. Used in concert, each method can help offset the weaknesses of the other while using their individual strengths to achieve the desired end result: accurate short- and long-term forecasts.

Weather and the Government

The National Weather Service has its roots in a February 9, 1870, joint resolution of Congress that authorized the secretary of war to establish a national weather service as a function of the U.S. Army Signal Corps.

The word *tornado* was banned from forecasts until 1938, because of the fear of public panic.

The Weather Bureau Severe Weather Unit, founded in 1952, was succeeded by the Severe Local Storm Warning Center in 1953 and renamed the National Severe Storms Forecast Center in 1965. Since 1982, it has been called the Storm Prediction Center and is now a part of the National Weather Center in Norman, Oklahoma.

Congress authorized funds for the U.S. Weather Bureau to create the National Hurricane Research Project in 1955.

The Air Force Hurricane Hunters of the 53rd Weather Reconnaissance Squadron fly into hurricanes to gather data. This practice began in 1944, when two air force pilots dared each other to attempt what had never before been done.

Seasons and Synoptics

In meteorological terms, significant weather—heat, cold, thunderstorms, tornadoes, etc.—begins with the synoptic, or big-picture, scale. When synoptic conditions—such as deep ridges and troughs and the jet stream—are strong, weather extremes often occur: heat waves, subzero temperatures, severe thunderstorms, ice storms, and blizzards. Without the synoptic influence, weather is unremarkable: seasonal temperatures, fair skies or scattered clouds, and brief showers.

Weather travels in waves around the globe, repeating the cycle of high-pressure ridge to low-pressure trough back to high-pressure ridge over and over, as mid- and upper-level winds keep the atmosphere in perpetual motion. In effect, this is the atmosphere performing a balancing act in its attempt to equalize pressure. A high-pressure ridge (H on a weather map) is noted for fair weather that is either hot in the warm seasons or chilly in the cold seasons. Low pressure (L on a weather map) is associated with storms in any season, and the deeper the trough, the more severe the storm potential. A cold or stationary front is linked to the low pressure, and it is along this front that weather events occur. Weather fronts, which are near the Earth's surface, usually advance along a north or northeasterly to south or southeasterly path.

The synoptic-scale equivalent in astrometeorology is the seasonal ingress, or the beginning of each season, when the Sun enters a cardinal sign (Aries, spring; Cancer, summer; Libra, autumn; Capricorn, winter). Based on the planetary positions at the time of the ingress, astrometeorology gives insight into the seasonal weather trends in any location. For example, if Saturn influences a particular region, cool, cloudy days will outnumber sunny, warm ones, and there will be an above-average number of storms during the season's three-month period. When Neptune is active in a particular location at the ingress, excessive rainfall and increased potential for flooding are indicated.

At best, meteorologists can predict weather with accuracy up to about three days in advance. They track ridges and troughs, among other factors, to formulate these forecasts, predicting how fast and into what location a trough or ridge will advance. If you live on the East Coast, for example, look to the west to see the weather that might arrive in your location in a few days, depending upon its exact easterly path.

Meteorologists are unable to predict weeks, months, or years in advance what the general weather conditions will be during a season in a particular location. (Meteorologists do issue seasonal forecasts, but this is a complex area of meteorology that requires a high level of expertise, and more advanced technology is needed to increase its accuracy.)

Astrometeorology, on the other hand, can be used to predict not only this winter's or next summer's seasonal weather but also next week, next weekend, next month, or ten or fifty years in the future. When used in conjunction with traditional meteorology, the seasonal trends indicated by astrometeorology can be used to pinpoint where a ridge or trough (and its attendant frontal system) are most likely to occur. For example, a deep trough off the coast in the northwest Pacific could move eastward over the northern tier of states or farther south across the Plains and into the lower Mississippi River Valley, reaching either of these regions in five to seven days. By using astrometeorology and the planetary placements at the ingress, it's possible to predict whether the path of the storm (and trough) is likely to be more northerly or southeasterly.

Zodiacal sign placements at the seasonal ingress also give the astrometeorologist information regarding temperature and moisture. Fire signs (Aries, Leo, Sagittarius), for example, indicate higher temperatures and dryness, while water signs (Cancer, Scorpio, Pisces) reflect moisture levels ranging from normal to above. Here, again, a combination of meteorology and astrometeorology can be used to refine seasonal forecasts and zero in on areas where specific weather patterns will develop, including weeks when the jet stream, for example, will dip far south and create icy conditions.

Lunar Phases and the Mesoscale

The mesoscale is where the action is in weather forecasting—thunderstorms, winter storms, tornadoes, and squalls. Mesoscale weather is short-term, from a few hours to a day, when a front moves through an area bringing severe weather, heavy precipitation, and high winds. This is the daily forecast as well as the watches and warnings issued by the Storm Prediction Center. To forecast thunderstorms, for example, meteorologists use radar and many other computer-based tools to measure temperature, dew point and relative humidity (moisture), and wind, all of which are factors in determining whether convection (thunderstorms) will occur.

Weather, however, doesn't just "happen" at a given location. Instead, conditions favorable for convection (and other weather phenomena) begin to develop west or north of where a thunderstorm actually occurs. These are the factors meteorologists assess prior to issuing the day's forecast, including where to expect weather to develop and the day's high and low temperatures. Weather extremes usually occur when the short-term mesoscale factors interact with long-term, synoptic-scale factors. When conditions favor convection and there is also a deep trough in the area, morning meteorologists might forecast severe thunderstorms and an increased possibility of tornadoes. But on many days, it is only after convection is well underway that meteorologists can forecast the chance for hail or tornadoes. Mpst meteorological forecasts, watches, and warnings usually also cover a multistate area.

The astrometeorology equivalent of mesoscale forecasting is the lunar phase. (Four lunar phases occur in every twenty-eight-day period: New Moon, first quarter Moon, Full Moon, and fourth quarter Moon.) By using a chart calculated for each of these phases, which are in effect for approximately seven days, the astrometeorologist can identify areas prone to severe weather as well as temperature, moisture, and wind. These weekly trends, like the seasonal trends forecasted from the ingress chart, are determined based on planetary placements and zodiac signs in a given location. As is true with meteorology and the synoptic-mesoscale connection, the most significant weather events occur when lunar phase alignments strongly connect with that season's ingress alignments.

Just as the meteorologist looks for certain atmospheric conditions that favor severe thunderstorms and tornadoes, the astrometeorologist is alert for planetary signatures, such as Mars-Uranus (heat-cold) contacts, which reflect the temperature contrast necessary for deep convection and an increased possibility for tornadoes. Astrometeorology can also narrow the zone for tornado formation well ahead of the actual occurrence; because of the limitations of radar, meteorologists generally can do this only minutes before a funnel cloud descends from a supercell thunderstorm.

Tropical Forecasting

Tropical storms and hurricanes have the components of both the synoptic scale and mesoscale, both ingress and lunar phase.

The hurricane itself is a synoptic-ingress event, while the severe thunderstorms and tornadoes that occur in the eyewall (area immediately surrounding the hurricane's eye) are mesoscale-lunar phase events.

Computer modeling has not been perfected to the point where meteorologists can determine far in advance where a hurricane will make landfall. Its path depends in large part upon mid- and upper-level winds that steer the hurricane; these winds shift direction due to the influence of other, non-hurricane atmospheric influences. At times, computer models forecast vastly different tracks for the same hurricane, with paths that range, for example, from northward along the Florida coast to straight westward toward Central America.

Astrometeorology does not have this limitation. In fact, the astrometeorologist can pinpoint the exact location a hurricane will make landfall by analyzing planetary alignments in the ingress and lunar phase charts and correlating them to longitude and latitude.

Meteorology or Astrometeorology?

The best option by far is to use both. But science-oriented meteorologists are skeptical of astrology-oriented astrometeorologists, and most astrometeorologists have only a smattering of meteorology knowledge. Further research will advance both fields, and cooperative efforts will maximize the potential of both for the public's benefit.

About the Author

Kris Brandt Riske is executive director and a professional member of the American Federation of Astrologers, the oldest U.S. astrological organization, founded in 1938; and a member of NCGR (National Council for Geocosmic Research). She has a master's degree in journalism and a Certificate of Achievement in Weather Forecasting from Penn State. Kris is the author of several books, including: Llewellyn's Complete Book of Astrology: The Easy Way to Learn Astrology; Mapping your Money; Mapping Your Future; Astrometeorology: Planetary Powers in Weather Forecasting; *and coauthor of* Mapping Your Travels and Relocation. *She also writes for astrology publications, and does the annual weather forecast for* Llewellyn's Moon Sign Book. *In addition to astrometeorology, she specializes in predictive astrology. Kris is an avid NASCAR fan, although she'd rather be a driver than a spectator. She also enjoys gardening, reading, jazz, and her three cats.*

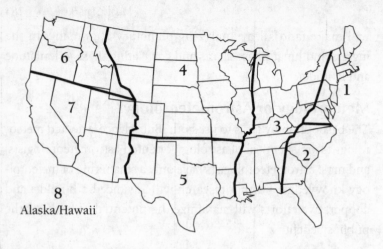

8 Alaska/Hawaii

Weather Forecast for 2010

By Kris Brandt Riske

Winter 2010

Zones 1 and 2 will be windy this season, with temperatures seasonal to below and precipitation levels ranging from average to below. Both Saturn and Pluto influence zone 3, indicating a tendency for cloudy skies and a lower range of temperatures and precipitation; this zone is also prone to major winter storms.

Eastern areas of zones 4 and 5 are more likely to see near-normal precipitation levels because of Venus' influence, but Uranus here also indicates a strong tendency for cold, dry weather. The other areas in these zones will be dry with temperatures ranging from seasonal to above; however, because of the influence of Jupiter and Neptune, areas from eastern Wyoming through western North Dakota can expect abundant precipitation at times.

Precipitation in zones 6 and 7 will range from average to below in both temperature and precipitation, but Arizona, eastern Utah, and western Colorado will see periods of abundant downfall, as will northern coastal areas. Locations from southern California to Nevada, Idaho, and Montana will be generally dry but also prone to major storms.

Eastern and central Alaska will be dry this season, with more precipitation in eastern parts of the state; temperatures will be seasonal. Hawaii can expect temperatures ranging from seasonal to above, with average precipitation.

January 1–6, Full Moon

Zone 1: The zone is windy with temperatures seasonal to below, and southern areas have a chance for precipitation.

Zone 2: Weather is fair and seasonal north; central and southern areas see precipitation, with potential for a winter storm and severe thunderstorms.

Zone 3: Much of the zone is fair and seasonal, with some cloudiness east and scattered precipitation.

Zone 4: High winds accompany precipitation in western and central areas, and eastern parts of the zone are mostly fair; temperatures range from seasonal to below.

Zone 5: Wind and precipitation occur in western and central areas, and eastern areas see variable cloud cover; temperatures are seasonal to below.

Zone 6: Overcast skies accompany stormy conditions across much of the zone, with high winds signaling blizzard potential in some areas.

Zone 7: The zone is overcast, cold, and windy with the highest precipitation in coastal areas as a low-pressure system advances across much of the zone; temperatures are more seasonal in eastern areas.

Zone 8: Much of Alaska is stormy and windy with abundant downfall in some areas; western parts of the state are generally fair. Hawaii is cloudy, wet, and windy with temperatures seasonal to below.

January 7–13, 3rd Quarter Moon

Zone 1: The zone is windy with precipitation and seasonal temperatures.

Zone 2: Central and southern areas are cool, cloudy, and windy with scattered precipitation, and northern skies are variably cloudy.

Zone 3: Western and central areas of the zone are overcast and windy with precipitation and blizzard potential, while eastern areas are fair and very windy.

Zone 4: Western areas see precipitation, and the Plains and eastern parts of the zone are very cold, dry, and windy under a high-pressure system.

Zone 5: Western areas are windy and fair; northern areas in the central part of the zone see scattered precipitation, but these states are mostly cold, fair, and dry as polar air dips south; eastern areas are not as cold, with scattered clouds and wind.

Zone 6: Western and central areas see abundant precipitation under stormy skies that extend across the zone; temperatures are more seasonal east.

Zone 7: A winter storm brings abundant precipitation to coastal and mountain areas of the zone, and eastern areas see scattered precipitation later in the week; temperatures are cooler west and central.

Zone 8: Alaska is stormy in the west with abundant precipitation that moves eastward into central areas of the state; fair and windy east. In Hawaii, precipitation brings cooler temperatures west, while central and eastern areas are initially fair with increasing clouds and precipitation later in the week.

January 14–22, New Moon

Zone 1: Northern skies are fair, and southern areas are cloudy with precipitation, some possibly abundant.

Zone 2: Northern areas see precipitation, while central and southern areas are fair to partly cloudy and very windy.

Zone 3: Much of the zone is cold, dry and fair; central and eastern areas are windy, with eastern areas seeing precipitation, some abundant; temperatures are seasonal to below.

Zone 4: Western areas through the central Plains see precipitation followed by colder temperatures, and eastern areas are fair and cold.

Zone 5: Precipitation prevails as a front moves across the zone, but eastern areas are mostly fair to partly cloudy; temperatures are seasonal to below.

Zone 6: Cloudy, stormy western skies yield abundant precipitation, while central areas are variably cloudy, and eastern areas fair; temperatures are seasonal to below.

Zone 7: Eastern areas are fair to partly cloud and seasonal under a high-pressure system, central parts of the zone see variable cloudiness, northern coastal areas are wet and some areas see abundant precipitation, and southern coastal areas are fair to partly cloudy.

Zone 8: Eastern Alaska is cold with variable cloudiness and precipitation, central areas are seasonal with scattered precipitation, and western areas are fair to partly cloudy. Hawaii sees scattered precipitation under variably cloudy skies, temperatures are seasonal, and eastern areas are windy.

January 23–29, 1st Quarter Moon

Zone 1: The zone is variably cloudy and cold, with precipitation south and fair to partly cloudy skies north.

Zone 2: Temperatures are cold in northern areas with scattered precipitation, and southern and central areas are more seasonal and fair to partly cloudy.

Zone 3: Fair skies prevail to the west, with more cloudiness in central and east; eastern areas see precipitation, followed by colder temperatures; temperatures across the zone are seasonal to below.

Zone 4: Much of the zone is variably cloudy and cold with scattered precipitation, while the eastern Plains are windy and more seasonal.

Zone 5: Variable cloudiness prevails across the zone, while central areas are cold; central and eastern areas see precipitation, some locally heavy as a front advances.

Zone 6: Western areas are windy and cool with a chance for precipitation, central areas are mostly fair and seasonal, and eastern areas are windy and cold with scattered precipitation.

Zone 7: Western and central areas are fair and windy, eastern areas see variable cloudiness and scattered precipitation, and temperatures are seasonal to above.

Zone 8: Alaska is stormy, cold, and windy central and east with precipitation, and western skies are mostly fair with seasonal temperatures. Western areas of Hawaii are fair and breezy, while central and eastern areas see showers and thunderstorms.

January 30–February 4, Full Moon

Zone 1: Temperatures are seasonal to below with variably cloudy skies, and windy south.

Zone 2: Skies are fair to partly cloudy north, with precipitation central and south, possibly freezing rain or sleet; temperatures are seasonal to below.

Zone 3: Western areas are windy with precipitation that advances into central parts of the zone, where temperatures are colder; eastern areas are fair to partly cloudy and seasonal.

Zone 4: Temperatures are seasonal to below across the zone with precipitation east.

Zone 5: Skies are mostly fair with variable cloudiness east; temperatures are seasonal to below.

Zone 6: Western areas are cold and windy with a chance for precipitation; central areas are cold, becoming more seasonal; eastern areas see precipitation and colder temperatures as a front moves through.

Zone 7: Variable cloudiness and temperatures ranging from seasonal to below prevail in western and central areas, which see precipitation; eastern areas are fair and windy, becoming colder.

Zone 8: Western and central Alaska are fair to partly cloudy, and eastern areas are windy with precipitation and temperatures seasonal to below. Hawaii is fair with temperatures seasonal to above; western and central areas are windy with scattered precipitation.

February 5–12, 4th Quarter Moon

Zone 1: The zone is variably cloudy, and windy south; temperatures are seasonal to below.

Zone 2: Temperatures are seasonal to below with abundant precipitation in northern areas.

Zone 3: Western areas are cold and windy with precipitation that advances across the zone, bringing abundant downfall to eastern areas.

Zone 4: The zone is mostly fair and seasonal west, colder central and east; central areas are very windy and cloudy, and eastern areas are fair to partly cloudy.

Zone 5: Eastern areas see scattered precipitation and variable cloudiness, while western areas are fair to partly cloudy, and central parts of the zone are windy and cloudy; temperatures are seasonal to below.

Zone 6: The zone is generally fair to partly cloudy with scattered precipitation west and clouds increasing in eastern areas with a chance for precipitation; temperatures are seasonal to below.

Zone 7: Scattered precipitation is possible, but the zone is mostly fair and seasonal.

Zone 8: Western Alaska sees abundant precipitation, central and eastern areas are fair, and temperatures are seasonal to below. Hawaii is fair and seasonal with scattered precipitation in western areas later in the week.

February 13–20, New Moon

Zone 1: Much of the zone sees scattered precipitation, with seasonal temperatures south, and clearing skies and colder temperatures north.

Zone 2: The zone is variably cloudy and seasonal with scattered precipitation.

Zone 3: Western areas are very cold, overcast, and windy, and eastern areas are more seasonal with precipitation.

Zone 4: Western areas see precipitation, and much of the zone is cold, with temperatures dipping in the eastern Plains under overcast skies.

Zone 5: Temperatures are below normal and much colder east and the zone is windy; western areas see precipitation, which advances into central parts of the zone.

Zone 6: Precipitation advances from western to central areas, temperatures are seasonal, and much of the zone is windy.

Zone 7: Southern coastal areas are initially fair with increasing cloudiness, northern coastal and central areas see precipitation, and eastern areas are partly cloudy; temperatures are seasonal to below.

Zone 8: Central Alaska is stormy, western areas are cold with precipitation later in the week, and eastern areas are fair and seasonal. Eastern Hawaii is very windy with precipitation, western areas of the state are fair, and central areas are variably cloudy with scattered precipitation.

February 21–27, 1st Quarter Moon

Zone 1: Temperatures dip under overcast skies that bring precipitation to much of the zone.

Zone 2: Temperatures are seasonal to below, and clouds increase in northern areas, bringing precipitation, some abundant; central and southern areas are variably cloudy.

Zone 3: Temperatures are seasonal but colder east, and much of the zone sees precipitation.

Zone 4: Northwestern areas see precipitation, western areas and western Plains states are windy under variably cloudy skies, eastern Plains are fair, and the Mississippi River Valley sees precipitation; temperatures are seasonal but colder west.

Zone 5: The zone is fair to partly cloudy and seasonal.

Zone 6: Western areas are mostly fair and seasonal with a chance for precipitation; central areas are windy, cloudy, and cold, with precipitation moving east.

Zone 7: Southern coastal areas are windy and cloudy with precipitation that advances into central and eastern areas, with the heaviest downfall east; northern coastal areas are fair and windy; temperatures are seasonal to below.

Zone 8: Central and eastern Alaska are cold and stormy as a front moves eastward, and western areas are fair. Much of Hawaii sees showers, and western areas are cool, cloudy, and windy.

February 28–March 6, Full Moon

Zone 1: Temperatures are seasonal to below under variably cloudy skies, with windy conditions and precipitation south.

Zone 2: The zone is fair to partly cloudy with temperatures ranging from seasonal to above; southern areas are windy, and northern areas see scattered precipitation.

Zone 3: Western skies are partly cloudy, and eastern areas see precipitation, some abundant; temperatures are seasonal to above.

Zone 4: Conditions are windy and partly cloudy west, and most of the zone is dry with variable clouds and temperatures seasonal to above.

Zone 5: Variable cloudiness prevails across the zone, which is dry with temperatures ranging from seasonal to above.

Zone 6: The zone is seasonal with a chance of precipitation east.

Zone 7: The zone sees scattered precipitation, variable cloudiness, and seasonal temperatures.

Zone 8: Eastern Alaska sees precipitation, and central and eastern areas are mostly fair and seasonal. Much of Hawaii is fair and seasonal, but it is windy east and partly cloudy.

March 7–14, 3rd Quarter Moon

Zone 1: Fair to partly cloudy skies prevail, with more cloudiness and scattered precipitation north; temperatures are seasonal.

Zone 2: The zone is variably cloudy, northern areas are dry, and central and southern areas see precipitation.

Zone 3: The zone is fair to partly cloudy with more clouds east; temperatures are seasonal to above.

Zone 4: Fair skies prevail to the west, and central and eastern areas see high winds and thunderstorms, some severe; temperatures are seasonal to above.

Zone 5: Temperatures are seasonal to above, eastern areas see scattered thunderstorms, and central and western parts of the zone are fair.

Zone 6: Windy weather and precipitation in western areas advances into central parts of the zone; eastern areas are fair and seasonal.

Zone 7: The zone is windy and fair except for a chance of precipitation in southern coastal and central parts of the zone.

Zone 8: Alaska is mostly fair and seasonal with precipitation and cooler temperatures east. Eastern areas of Hawaii are cooler with precipitation, and central and western areas are mostly fair.

March 15–22, New Moon

Zone 1: Southern areas, which see precipitation, are cloudier than northern parts of the zone; temperatures are seasonal to below.

Zone 2: Much of the zone is stormy with high winds and the potential for abundant rainfall north and severe thunderstorms central and south.

Zone 3: Western areas are fair and seasonal, and central and eastern areas see more cloudiness and cooler temperatures with precipitation east.

Zone 4: The western Plains are very windy, and the zone is generally dry with temperatures ranging from seasonal to above.

Zone 5: Fair skies prevail across the zone, which is windy in central areas with temperatures ranging from seasonal to above; severe thunderstorms are possible.

Zone 6: The zone is windy, cold, and wet with major storm potential throughout.

Zone 7: Southern coastal areas are variably cloudy with a chance for precipitation, northern coastal areas are very windy and cold with precipitation, eastern areas are windy and dry with a chance for precipitation at the end of the week; temperatures are mostly seasonal to above.

Zone 8: Fair skies and seasonal temperatures prevail in Alaska, with scattered precipitation west. Hawaiian skies are generally sunny and temperatures seasonal, but clouds prevail in western areas.

Spring 2010

Zones 1 and 2 and the northeastern part of zone 3 will be cold, with an above-average number of major storms because of the influence of Mercury and Pluto. Zone 3, although not as cold, will nevertheless experience temperatures below normal, cloudy skies, and heavy downfall at times.

Weather will be generally dry in the eastern and western parts of zones 4 and 5 this spring, with temperatures seasonal to above, and central areas in these zones will see temperatures ranging from normal to below normal and a higher potential for strong cold fronts as well as cloudy skies.

Temperatures in zones 6 and 7 will be average to above, with conditions ranging from normal to dry. Areas from western Montana south through central Arizona, under the influence of Neptune, are prone to periods of abundant precipitation and flooding.

Alaska will be generally seasonal with average precipitation this spring, as will Hawaii.

March 23–28, 1st Quarter Moon

Zone 1: Northern areas are windy, and southern areas are overcast and stormy with abundant precipitation; temperatures are seasonal to below.

Zone 2: Stormy, overcast skies prevail to the north, and central and southern areas are mostly fair with a chance of thunderstorms; temperatures are seasonal to below.

Zone 3: Western areas see scattered thunderstorms and temperatures ranging from seasonal to above, and eastern areas see abundant precipitation in some areas.

Zone 4: The central and eastern Plains see scattered thunderstorms, but skies are generally fair, conditions dry, and temperatures seasonal to above; western areas are cloudy.

Zone 5: The zone is windy, with cloudy skies west, and mostly fair skies central and east with scattered thunderstorms east; temperatures are seasonal to above.

Zone 6: Western areas are fair to partly cloudy, central areas see scattered precipitation, and eastern areas are windy; temperatures are seasonal to below.

Zone 7: Northern coastal areas are cloudy and cool with precipitation, and central and eastern areas are windy with variable cloudiness; temperatures are seasonal to below.

Zone 8: Precipitation prevails across Alaska, with abundant downfall west, followed by clearing and fair skies; temperatures are seasonal. Western Hawaii is mostly fair and seasonal, and temperatures are cooler central and east.

March 28–April 5, Full Moon
Zone 1: Southern areas see scattered precipitation and windy conditions, while skies are partly cloudy north; temperatures are seasonal to below.

Zone 2: Central and southern areas are mostly fair, seasonal, and dry, and northern areas see variable cloudiness and scattered precipitation.

Zone 3: The zone is seasonal with variable cloudiness and scattered precipitation.

Zone 4: Much of the zone sees precipitation with thunderstorms and windy conditions as a front advances; western skies are cloudier; temperatures are seasonal.

Zone 5: Western areas see precipitation, while severe thunderstorms are possible in central parts of the zone; eastern areas see precipitation later in the week; temperatures are seasonal.

Zone 6: The zone is partly cloudy and windy with scattered precipitation, but western areas are fair; temperatures are seasonal to above.

Zone 7: Fair skies and temperatures ranging from seasonal to above prevail across the zone, with a chance for precipitation.

Zone 8: Alaska sees scattered clouds with temperatures seasonal to above. Hawaii is fair and seasonal.

April 6–13, 3rd Quarter Moon

Zone 1: The zone is cloudy and cool with precipitation, some abundant.

Zone 2: Much of the zone sees precipitation and thunderstorms, some severe central and south; temperatures are seasonal to above.

Zone 3: Western areas are fair and dry, and central parts of the zone see thunderstorms, some severe, while eastern areas see scattered precipitation; temperatures are seasonal to above.

Zone 4: The zone is mostly fair and dry with temperatures ranging from seasonal to above, and western areas see scattered precipitation.

Zone 5: Temperatures are seasonal to above with variable cloudiness west and central and a chance for scattered thunderstorms.

Zone 6: The zone is mostly cloudy and windy with precipitation, some abundant in western areas; eastern areas are fair and seasonal.

Zone 7: Northern coastal areas see precipitation that advances into central parts of the zone; southern coastal and eastern areas are mostly fair; and temperatures are seasonal but hot in the desert.

Zone 8: Western Alaska is stormy, and central and eastern areas are fair and seasonal. Much of Hawaii is fair, with western skies becoming cloudy, bringing precipitation.

April 14–20, New Moon

Zone 1: The zone is cloudy with precipitation, windy south, and temperatures seasonal to below.

Zone 2: Northern areas see precipitation, and central and southern areas see scattered thunderstorms with high winds.

Zone 3: Western areas are variably cloudy and seasonal with a chance for thunderstorms, central areas see above normal temperatures and thunderstorms, and eastern areas are cooler with precipitation.

Zone 4: The zone is mostly fair with temperatures ranging from seasonal to above; eastern areas are cooler with scattered precipitation.

Zone 5: The zone is fair with above-normal temperatures, and eastern areas are seasonal with a chance for thunderstorms.

Zone 6: Stormy skies in western areas advance eastward with precipitation across the zone; temperatures are mostly seasonal.

Zone 7: Coastal and inland areas see precipitation and high winds with heaviest downfall north, and central and eastern areas are windy; the desert is fair with above-normal temperatures.

Zone 8: Alaska is fair and windy west, with precipitation west and central; temperatures are seasonal. Eastern and central Hawaii are cloudy with precipitation, and western skies are fair.

April 21–27, 1st Quarter Moon

Zone 1: Northern areas are cool and cloudy with locally heavy precipitation, and southern areas see scattered precipitation.

Zone 2: Northern areas are mostly fair, with some cloudiness central and south bringing precipitation; temperatures are seasonal to below.

Zone 3: High temperatures spark thunderstorms, some severe, in western areas; central areas are cooler with precipitation; and eastern areas are fair.

Zone 4: Much of the zone is fair with temperatures above normal and some potentially severe thunderstorms, especially in eastern areas.

Zone 5: The zone is partly cloudy with temperatures seasonal to above and a chance for thunderstorms east.

Zone 6: Western parts of the zone are overcast with precipitation that advances into central and eastern areas, bringing scattered precipitation and windy conditions.

Zone 7: Southern coastal, central, and desert areas see high temperatures, while northern coastal areas are cool and cloudy with showers and seasonal temperatures.

Zone 8: Central Alaska sees precipitation, some abundant, which moves eastward; western areas are fair. Hawaii is humid and cloudy with showers, some locally heavy.

April 28–May 5, Full Moon

Zone 1: The zone is mostly fair but with more cloudiness north, where locally heavy precipitation and strong thunderstorms are possible; temperatures are seasonal.

Zone 2: Northern areas are fair, and central and southern areas are cloudy with potential for strong thunderstorms; temperatures are seasonal to above.

Zone 3: Western areas see precipitation and thunderstorms, some severe with tornado potential; central areas have a chance for precipitation; eastern areas are sunny; and temperatures are seasonal to above.

Zone 4: Skies are mostly sunny west with a chance for precipitation later in the week; the western Plains are seasonal; and temperatures rise in the eastern Plains and Mississippi River Valley, which see precipitation and scattered thunderstorms, some severe with tornado potential.

Zone 5: Skies are fair east and west, and cloudy central with scattered thunderstorms.

Zone 6: Eastern areas see precipitation, and western and central areas are variably cloudy and windy; temperatures range from seasonal to above.

Zone 7: The zone is variably cloudy, with more cloudiness east with a chance for showers; temperatures are seasonal.

Zone 8: Most of Alaska is fair and seasonal, with precipitation in central areas. Hawaii sees showers and temperatures are seasonal but warmer west.

May 6–12, 3rd Quarter Moon

Zone 1: Seasonal temperatures and scattered showers prevail across the zone, with some northern areas seeing more downfall; southern areas are very windy.

Zone 2: Northern areas are very windy, and central and southern areas are humid with scattered thunderstorms, some severe.

Zone 3: Severe thunderstorms with tornado potential are possible in western areas, and central areas see showers; temperatures are above normal, but cooler east.

Zone 4: The zone is generally sunny and dry with high winds and severe thunderstorms in some areas as temperatures rise.

Zone 5: The zone is humid and generally sunny with high temperatures and scattered thunderstorms, some severe with tornado potential.

Zone 6: Western and central areas see showers, some bringing abundant downfall as a front advances, followed by cooler temperatures; temperatures are seasonal to below.

Zone 7: Temperatures are seasonal to above, eastern areas are cloudy, and western and central parts of the zone see scattered showers.

Zone 8: Alaska is variably cloudy and seasonal, with precipitation in central areas. Eastern Hawaii see showers and thunderstorms, some strong, and then cooler temperatures; western and central Hawaii is fair and seasonal.

May 13–19, New Moon

Zone 1: Cloudy skies prevail across the zone with potential for abundant precipitation north, where temperatures are cooler.

Zone 2: Northern areas see showers, and central and southern areas are fair and seasonal.

Zone 3: Much of the zone is fair, but eastern areas see showers and scattered thunderstorms; temperatures are seasonal.

Zone 4: Western areas are fair and dry under high temperatures; central areas see precipitation, some abundant, with severe thunderstorm and tornado potential; and eastern areas are fair, windy, and seasonal.

Zone 5: Dryness and high temperatures prevail to the west, with scattered thunderstorms; severe thunderstorms and tornadoes are possible in central areas of the zone; and week's end could bring abundant precipitation and flooding east.

Zone 6: Some western areas see locally heavy precipitation, and the rest of the zone is variably cloudy and windy with scattered precipitation.

Zone 7: Much of the zone is fair, windy, and dry under a high-pressure system, and northern coastal areas see showers.

Zone 8: Western and central Alaska is cold and cloudy with abundant precipitation in some locations, and eastern areas are mostly fair; temperatures are seasonal to below. Hawaii is windy with scattered precipitation, abundant in some areas.

May 20–26, 1st Quarter Moon

Zone 1: The zone is fair and seasonal.

Zone 2: Northern areas are fair and seasonal, and abundant precipitation is possible central and south with strong thunderstorms and tornadoes.

Zone 3: Central and eastern areas of the zone are mostly fair, windy, and seasonal, and western parts of the zone see high temperatures with thunderstorms, some severe.

Zone 4: Much of the zone is windy and dry under a high-pressure system, with scattered thunderstorms east, some severe; temperatures are seasonal to above.

Zone 5: Western areas are windy with variable cloudiness, and central and eastern areas see scattered thunderstorms, some possibly severe.

Zone 6: The zone is fair with temperatures seasonal to below and scattered precipitation central and east.

Zone 7: Fair weather prevails west, skies are variably cloudy central and east with a chance for precipitation north, and temperatures are seasonal to above.

Zone 8: Most of Alaska is fair with temperatures seasonal to below, and eastern areas are windy with precipitation. Hawaii is sunny and seasonal.

May 27–June 3, Full Moon

Zone 1: The zone is fair, windy, and seasonal.

Zone 2: Northern skies are fair, while southern and central areas have the potential for severe thunderstorms, tornadoes, and abundant precipitation.

Zone 3: Eastern areas are fair with scattered precipitation, and the rest of the zone has the potential for severe thunderstorms and tornadoes.

Zone 4: Much of the zone is windy, western and eastern areas are fair, and the Plains have the potential for severe thunderstorms and tornadoes.

Zone 5: Central and eastern areas see scattered thunderstorms, some severe, and western areas are fair; temperatures are seasonal.

Zone 6: Western areas see showers, central areas are fair, and eastern skies are variably cloudy with scattered showers; temperatures are seasonal to below.

Zone 7: Sunny skies prevail across most of the zone, with high clouds east; temperatures are seasonal to above.

Zone 8: The Alaskan zone is fair and seasonal. Hawaii is sunny and seasonal.

June 4–11, 3rd Quarter Moon

Zone 1: The zone is seasonal and humid with precipitation, some abundant, with flood potential.

Zone 2: Northern areas see significant precipitation and flood potential; central and southern parts of the zone are windy with scattered thunderstorms, some severe with tornado potential; and conditions are humid with temperatures seasonal to above.

Zone 3: Much of the zone sees thunderstorms, some severe with tornado potential, and abundant downfall east increases flood potential; humidity and temperatures rise across the zone.

Zone 4: Western parts of the zone see rising temperatures with high winds and precipitation, while eastern areas see more cloudiness with scattered thunderstorms, some severe with tornado potential, and the Plains are generally dry.

Zone 5: Temperatures are seasonal to above, western areas see scattered thunderstorms, and east areas have a chance for precipitation later in the week.

Zone 6: Precipitation, some abundant, prevails across the zone with a second front moving in later in the week.

Zone 7: Western and central areas are windy with precipitation later in the week, and eastern areas see more cloudiness and precipitation, followed by cool temperatures.

Zone 8: Alaska is mostly fair and seasonal, but with more cloudiness and precipitation east. Hawaii is sunny with temperatures seasonal to above.

June 12–17, New Moon

Zone 1: The zone is cloudy with precipitation, followed by cooler temperatures.

Zone 2: Central and southern areas are humid with scattered precipitation and a chance for strong thunderstorms, northern areas are fair and humid, and temperatures are seasonal to above.

Zone 3: The zone is dry and fair with a chance for precipitation central and east.

Zone 4: The zone is fair to partly cloudy and humid with temperatures seasonal to above; central areas of the zone have a chance for severe scattered thunderstorms.

Zone 5: The zone is fair, dry, and hot, with scattered thunderstorms east, some strong.

Zone 6: Fair skies prevail west with precipitation later in the week, central areas are partly cloudy with scattered thunderstorms followed by cooler temperatures, and eastern areas are under a high-pressure system.

Zone 7: The zone is generally fair and seasonal, with precipitation in northern coastal areas at week's end, scattered precipitation central, and rising temperatures east.

Zone 8: Central and western Alaska are windy and cloudy with precipitation, and eastern parts of the state are fair; temperatures are seasonal to below. Hawaii is fair, seasonal, and windy, with precipitation and possibly strong thunderstorms west.

Summer

Summer temperatures will range from normal to above throughout zones 1 and 2 and in northeastern areas of zone 3, all of which will see below average precipitation. Much of zone 3 will be dry, but also prone to severe thunderstorms and strong cold fronts; temperatures will be seasonal to below.

Eastern areas of zones 4 and 5 also will be prone to strong cold fronts and severe thunderstorms this season, and temperatures will range from average to below. Central parts of these zones will be dry, with temperatures seasonal to above, and the chance for severe thunderstorms. Eastern New Mexico and western Texas may see more precipitation.

Precipitation levels in zones 6 and 7 will range from average to dry, and states from western Washington to western Utah and eastern Arizona-western New Mexico will have an increased chance for thunderstorms. California will be dry.

Alaska's temperatures will be mostly seasonal, but precipitation will be below normal, and Hawaii will experience similar weather this summer.

June 19–25, 1st Quarter Moon

Zone 1: The zone is humid and seasonal with increasing clouds bringing showers, followed by cooler temperatures.

Zone 2: Northern areas see scattered showers, followed by cooler temperatures, and central and southern areas are fair, dry, and seasonal.

Zone 3: Eastern and central parts of the zone are hot and humid with scattered thunderstorms, some severe, and eastern areas are

windy and seasonal with showers later in the week.

Zone 4: The zone is humid and dry with variably cloudy skies temperatures seasonal to above, and a chance for scattered thunderstorms in the Plains.

Zone 5: The zone is variably cloudy, hot, and humid, with a chance for showers in central areas.

Zone 6: Much of the zone sees precipitation and temperatures seasonal to below.

Zone 7: Western and central parts of the zone have a chance for scattered showers, and eastern areas are under a high-pressure system; temperatures are seasonal to above.

Zone 8: Central Alaska is stormy and cold with abundant precipitation, and eastern and western areas are seasonal. Hawaii is cloudy with precipitation and temperatures seasonal to below.

June 26–July 3, Full Moon

Zone 1: Southern areas, which are hot and humid, see showers later in the week, and northern areas are seasonal and dry.

Zone 2: The zone is hot and humid with thunderstorms central and south, some severe with tornado potential; weather is more seasonal in northern areas.

Zone 3: Much of the zone is humid with precipitation, some abundant, with possible strong thunderstorms and tornado potential; temperatures are seasonal to above.

Zone 4: The zone is hot and dry but cloudy and cooler east with precipitation and high winds later in the week.

Zone 5: Hot, dry weather dominates across the zone.

Zone 6: Western and central areas are windy, fair, and seasonal, and eastern areas see rising temperatures.

Zone 7: Eastern areas are fair, western and central areas see scattered precipitation later in the week, and temperatures are seasonal to above.

Zone 8: Alaska is fair and seasonal. Hawaii is sunny and seasonal.

July 4–10, 3rd Quarter Moon

Zone 1: The zone is dry with temperatures seasonal to above.

Zone 2: Northern areas are fair and seasonal, and southern and central areas are humid and variably cloudy with precipitation, some possibly abundant.

Zone 3: Much of the zone is hot and humid, with scattered showers west and abundant precipitation possible in central areas; eastern areas are dry and fair with seasonal temperatures.

Zone 4: Temperatures are seasonal to above and the zone is generally dry with a chance for showers and some scattered thunderstorms east.

Zone 5: The zone is hot and humid with a chance for severe thunderstorms east.

Zone 6: Central areas are windy with variably cloudiness, scattered precipitation, and temperatures seasonal to above; eastern parts of the zone are cloudy and cool with showers; and western areas are mostly fair.

Zone 7: Western and central areas are mostly fair with temperatures seasonal to above, western parts of the zone are windy, and eastern areas are hot and humid with a chance for precipitation.

Zone 8: Most of Alaska is fair, seasonal, and windy, but eastern areas are stormy with high winds. Hawaii is dry and sunny with unseasonably warm temperatures.

July 11–17, New Moon

Zone 1: The zone is variably cloudy and seasonal with showers in the north.

Zone 2: Southern and central areas see scattered thunderstorms, northern areas are variably cloudy, and the zone is humid with temperatures seasonal to above.

Zone 3: Western and central areas have a chance for showers and scattered thunderstorms, some strong, and eastern areas are partly cloudy and humid.

Zone 4: The zone is fair to partly cloudy, dry, and windy with temperatures seasonal to above and a chance for scattered thunderstorms east.

Zone 5: Skies are fair to partly cloudy, hot, windy, humid, and dry.

Zone 6: The zone is fair to partly cloudy and dry with temperatures seasonal to above.

Zone 7: Temperatures are seasonal to above and the zone is dry.

Zone 8: Western Alaska sees precipitation, and eastern and central parts of the state are fair with temperatures seasonal to above. Hawaii is sunny and seasonal with scattered precipitation west later in the week.

July 18–24, 1st Quarter Moon

Zone 1: Northern areas have a chance for showers, while southern parts of the zone are windy and humid; temperatures are seasonal.

Zone 2: The zone is hot, humid, and dry.

Zone 3: Western and central parts of the zone see severe thunderstorms with tornado potential and temperatures seasonal to above, and eastern areas are fair to partly cloudy.

Zone 4: Eastern areas are cloudy and cool, and western and eastern parts of the zone are hot and dry with a chance for showers.

Zone 5: The zone is hot, dry, and fair with a chance for showers mid-week.

Zone 6: Increasing clouds bring showers to western and central areas as a front moves eastward, and eastern parts of the zone are fair and seasonal but cooler at higher elevations.

Zone 7: Eastern areas have a chance for showers and thunderstorms, but the zone is mostly fair and dry with temperatures ranging from seasonal to above.

Zone 8: Alaska is mostly fair with temperatures seasonal to above and showers east. Temperatures rise in Hawaii, and eastern areas are windy with a chance for showers.

July 25–August 2, Full Moon

Zone 1: Central and southern areas of the zone are mostly fair and seasonal, and northern areas see showers.

Zone 2: The zone is fair and dry with temperatures ranging from seasonal to above.

Zone 3: Western areas are windy with strong thunderstorms and tornado potential, and central and eastern areas are more seasonal and humid.

Zone 4: Temperatures are seasonal to above, western areas are variably cloudy with a chance for thunderstorms, and the Plains and eastern parts of the zone see thunderstorms, some severe with tornado potential.

Zone 5: Western parts of the zone see scattered thunderstorms, some strong, and central and eastern areas are mostly fair, hot, and humid with a chance for severe thunderstorms with tornado potential east.

Zone 6: Western parts of the zone are fair and seasonal, and central areas see increasing clouds as a front, which advances into eastern areas, brings showers followed by cooler temperatures.

Zone 7: Western and central parts of the zone are variably cloudy with showers, and eastern areas are fair with a chance for precipitation later in the week; temperatures are seasonal.

Zone 8: Eastern Alaska is fair, western parts of the state see precipitation, and central areas are stormy. Central and western Hawaii see showers, and eastern areas of the state are fair; temperatures are seasonal.

August 3–8, 3rd Quarter Moon

Zone 1: Abundant precipitation is possible in northern areas of the zone, which are cloudy and cool, and seasonal temperatures accompany fair skies south.

Zone 2: Central and southern areas are hot and humid with severe thunderstorms and tornado potential, followed by cooler temperatures; northern weather is fair and seasonal.

Zone 3: Temperatures are seasonal to above, and the zone is windy and fair with scattered showers and thunderstorms.

Zone 4: The zone is fair, dry, humid, seasonal, and windy in the northern Plains and eastern areas.

Zone 5: Eastern areas are windy, and the zone is humid and dry.

Zone 6: Central and eastern areas see heavy precipitation with high winds and overcast skies, and western parts of the zone are fair.

Zone 7: Southern coastal areas see showers, northern coastal areas are fair, central parts of the zone are cloudy with severe thunderstorms, and eastern areas are hot with strong thunderstorms.

Zone 8: Alaska is mostly fair with precipitation west. Hawaii is sunny and seasonal.

August 9–15, New Moon

Zone 1: The zone is variably cloudy with seasonal temperatures.

Zone 2: Wind and humidity accompany seasonal temperatures and fair to partly cloudy skies.

Zone 3: Central and eastern areas are mostly fair and seasonal, while western areas see severe thunderstorms with high winds and tornado potential, possibly from a tropical storm or hurricane.

Zone 4: The zone is fair and dry west, with severe thunderstorms with tornado potential in the eastern Plains and Mississippi River Valley.

Zone 5: Weather is fair and seasonal west, and partly cloudy and humid central and east with high winds and severe thunderstorms and tornado potential, possibly from a tropical storm or hurricane.

Zone 6: The zone is variably cloudy with temperatures seasonal to above and a chance for thunderstorms.

Zone 7: Hot weather accompanies scattered thunderstorms.

Zone 8: Alaska is fair to partly cloudy and cool. Hawaii is variably cloudy and seasonal with a chance for showers east.

August 16–23, 1st Quarter Moon

Zone 1: The zone is mostly fair, seasonal, and humid, and windy north.

Zone 2: Temperatures are seasonal to above under mostly fair skies with a chance for showers.

Zone 3: Eastern and central parts of the zone are mostly fair and dry, and western parts of the zone see showers and thunderstorms, some with abundant downfall and tornado potential; temperatures are seasonal to above.

Zone 4: Western areas are dry and hot, central parts of the zones have a chance for showers and thunderstorms, and eastern areas are hot and humid with the potential for severe thunderstorms and tornadoes.

Zone 5: Hot, humid weather increases the chance for severe thunderstorms with tornado potential across much of the zone; a tropical storm is possible.

Zone 6: The zone is variably cloudy and seasonal with a chance for precipitation east.

Zone 7: Seasonal temperatures accompany variably cloudy skies.

Zone 8: Eastern Alaska is fair, and western and central parts of the state see precipitation; temperatures are seasonal to below. Hawaii is sunny and seasonal with a chance for showers east.

August 24–31, Full Moon

Zone 1: The zone is seasonal and mostly fair with a chance for showers.

Zone 2: Weather is cooler north with a chance for showers, and dry south and central.

Zone 3: Variable cloudiness accompanies temperatures seasonal to below, with scattered showers east.

Zone 4: Temperatures are seasonal to below in the zone and skies are variably cloudy.

Zone 5: The zone is mostly fair and seasonal.

Zone 6: Temperatures seasonal to above accompany showers central and east and partly cloudy skies west.

Zone 7: Southern coastal areas see showers, northern coastal areas are windy and partly cloudy, central parts of the zone have scattered thunderstorms, and eastern skies are fair; temperatures are seasonal to above.

Zone 8: Eastern Alaska is fair, and western and central areas are windy with precipitation. Hawaii is windy and fair with temperatures seasonal to above.

September 1–7, 3rd Quarter Moon
Zone 1: The zone is fair and seasonal.
Zone 2: Scattered precipitation and variable cloudiness accompany scattered precipitation across the zone.
Zone 3: Temperatures are seasonal, and western areas see showers, while central and eastern parts of the zone are fair to partly cloudy.
Zone 4: Eastern areas have a chance for showers, and variable cloudiness accompanies temperatures seasonal to below.
Zone 5: The zone is cooler west and central under variably cloudy skies, with a chance for showers in eastern areas where temperatures are warmer.
Zone 6: Western showers advance into central parts of the zone, and temperatures are seasonal to below.
Zone 7: Northern coastal areas see showers, and temperatures ranging from seasonal to below and variable cloudiness prevail across the zone.
Zone 8: Alaska is fair with temperatures ranging from seasonal to below. Hawaii is sunny, with windy conditions and a chance for showers east.

September 8–14, New Moon
Zone 1: The zone is mostly fair with temperatures seasonal to above and a chance for showers.
Zone 2: Conditions are windy with variably cloudy skies and temperatures seasonal to above.
Zone 3: Western areas have a chance for thunderstorms, central and eastern areas are windy and variably cloudy, and the zone is humid with temperatures seasonal to above.
Zone 4: Central and eastern parts of the zone are variably cloudy,

western areas are fair, and the zone is seasonal to above.

Zone 5: The zone is variably cloudy with temperatures seasonal to above.

Zone 6: Much of the zone has a chance for showers under variably cloudy skies and temperatures seasonal to above.

Zone 7: Central and eastern parts of the zone are fair to partly cloudy with scattered showers and thunderstorms and temperatures seasonal to above; fine weather prevails west and central.

Zone 8: Alaska is mostly fair and seasonal with a chance for precipitation central. Hawaii is seasonal and fair to partly cloudy.

September 15–22, 1st Quarter Moon

Zone 1: The zone is fair north and cloudy south with temperatures seasonal to below.

Zone 2: Northern areas are cloudy and cool, and central and southern areas see precipitation, some abundant, possibly from a tropical storm.

Zone 3: Eastern areas are cloudy and cool; central parts of the zone see precipitation, some abundant, followed by cooler temperatures, possibly from a tropical storm; and western areas see scattered showers with temperatures seasonal to above.

Zone 4: Western parts of the zone are windy with a chance for scattered showers and thunderstorms, and the rest of the zone is fair to partly cloudy and seasonal.

Zone 5: The zone is windy with scattered thunderstorms east and temperatures seasonal to above.

Zone 6: Variable cloudiness with temperatures seasonal to above.

Zone 7: Western skies are cloudy, eastern areas have a chance for thunderstorms, central parts of the zone are partly cloudy, and wind accompanies temperatures seasonal to above.

Zone 8: Alaska is seasonal and windy west and central with scattered precipitation. Hawaii is sunny with temperatures seasonal to above.

Autumn

Temperatures in zone 1 will range from average to above this autumn, and conditions will be generally dry, while zone 2 will see mostly seasonal temperatures with precipitation ranging from average to below. Some severe thunderstorms will occur in the southeastern areas of zone 2 (Georgia, South Carolina).

Zone 3 temperatures will be seasonal to below, and precipitation will be below average except in central areas, where Neptune's influence increases the potential for significant downfall that could lead to flooding. States from Kentucky southward will have a greater chance for major storms.

The dominant planetary influences in eastern areas of zones 4 and 5 this autumn indicate temperatures seasonal to below with average to below-average precipitation. The eastern and central Plains areas of these zones will see temperatures from seasonal to above, as well as precipitation ranging from average to below. Temperatures in western parts of these zones will be seasonal to above with precipitation ranging from average to below.

The eastern mountain locations of zones 6 and 7 will see some significant storms, but conditions will be generally dry with seasonal temperatures. Western and central parts of these zones can expect windy conditions and average to below precipitation with seasonal temperatures.

Alaskan temperatures will range from seasonal to below, and precipitation will be average. Hawaii will be generally dry this fall with seasonal temperatures.

September 23–29, Full Moon

Zone 1: The zone is fair to partly cloudy and seasonal.

Zone 2: Scattered showers accompany variable cloudiness and seasonal temperatures.

Zone 3: Much of the zone is cloudy with temperatures seasonal to below.

Zone 4: Western areas see showers and thunderstorms, central and eastern areas are variably cloudy and cooler, with a chance for precipitation east.

Zone 5: Central and eastern areas are dry with variable cloudiness, and western parts of the zone see thunderstorms; temperatures are seasonal to above.

Zone 6: Western parts of the zone are fair and seasonal, and central and eastern areas are cooler with precipitation.

Zone 7: The zone is fair and dry west, and windy with variable cloudiness and chance for showers central and east; temperatures are seasonal to above.

Zone 8: Western and central Alaska see precipitation, some abundant, and eastern areas are fair; temperatures are seasonal to below. Central and eastern Hawaii are cooler, and western parts of the state are warmer and partly cloudy with scattered showers.

September 30–October 6, 3rd Quarter Moon

Zone 1: The zone is very windy with temperatures seasonal to below and precipitation, some abundant, north.

Zone 2: Northern areas see precipitation, and central and southern areas are generally fair with temperatures seasonal to above and scattered thunderstorms, some strong with tornado potential.

Zone 3: Western areas are variably cloudy with a chance for scattered thunderstorms, and central and eastern parts of the zone are fair and seasonal to above-normal temperatures.

Zone 4: Northwestern parts of the zone are fair, while areas to the south are cloudy with precipitation; the Plains are mostly fair and seasonal with scattered precipitation.

Zone 5: Variable cloudiness prevails west and central with scattered precipitation, and eastern parts of the zone are mostly fair and seasonal.

Zone 6: Much of the zone is windy with precipitation as a front advances under cloudy skies, bringing abundant downfall to some areas.

Zone 7: The zone is variably cloudy, windy west and central, with temperatures seasonal to above; northern coastal areas are fair, and eastern areas see precipitation.

Zone 8: Much of Alaska is cold with precipitation, some abundant, but conditions are more seasonal west. Hawaii is very windy and wet, with some areas receiving abundant precipitation; a typhoon or tropical storm is possible.

October 7–13, New Moon

Zone 1: The zone is windy and temperatures are seasonal to below, with partly cloudy skies north and precipitation south.

Zone 2: Northern areas are windy with precipitation, and southern central parts of the zone are partly cloudy with scattered precipitation.

Zone 3: Much of the zone is fair to partly cloudy and seasonal, and eastern areas are windy with precipitation and then colder.

Zone 4: The zone is fair to partly cloudy and seasonal with a chance for precipitation in the Plains.

Zone 5: Western areas have a chance for precipitation, and the rest of the zone is fair to partly cloudy and seasonal.

Zone 6: Western and central parts of the zone are windy with precipitation, and eastern areas are fair; temperatures are seasonal to below.

Zone 7: Much of the zone sees precipitation, some abundant, under cloudy skies with temperatures seasonal to below.

Zone 8: Western and eastern parts of Alaska are fair, and central areas of the state see precipitation, some abundant; temperatures are seasonal to below. Hawaii is windy and fair with a chance for thunderstorms, and temperatures are seasonal to above.

October 14–21, 1st Quarter Moon

Zone 1: Much of the zone sees scattered precipitation followed by cooler temperatures.

Zone 2: Southern and central parts of the zone see showers and scattered thunderstorms, some possibly strong with tornado potential, and temperatures are seasonal to above.

Zone 3: Western and central parts of the zone are fair to partly cloudy, eastern areas see precipitation, and temperatures are seasonal to above.

Zone 4: The zone is seasonal and variably cloudy.

Zone 5: Seasonal temperatures and windy accompany fair to partly cloudy skies.

Zone 6: The zone is windy with precipitation, some abundant, as a front advances, and temperatures are seasonal to below.

Zone 7: Eastern areas are cloudy with precipitation, some abundant, and western and central parts of the zone are fair; northern coastal areas see precipitation.

Zone 8: Alaska is seasonal and windy with variable cloudiness. Hawaii is seasonal and fair east, with cloudy skies and precipitation west.

October 22–29, Full Moon

Zone 1: The zone is windy, fair to partly cloudy, and seasonal.

Zone 2: Seasonal temperatures accompany fair to partly cloudy skies across the zone.

Zone 3: Skies are variably cloudy with temperatures seasonal to above and scattered precipitation west.

Zone 4: Western areas see precipitation and the zone is variably cloudy with temperatures seasonal to above.

Zone 5: Central and eastern parts of the zone are fair to partly cloudy, western areas see precipitation, and temperatures are seasonal to above.

Zone 6: The zone is fair west, windy and fair east, and temperatures are seasonal with a chance for precipitation in central areas.

Zone 7: The zone is fair to partly cloudy, windy east with a chance for precipitation, and temperatures are seasonal to above.

Zone 8: Alaska is mostly fair and seasonal west and east, with precipitation central. Hawaii is windy and seasonal, with precipitation east.

October 30–November 5, 3rd Quarter Moon

Zone 1: The zone is fair to partly cloudy with temperatures seasonal to above and windy south.

Zone 2: Northern areas are windy, and the zone is variably cloudy with temperatures seasonal to above.

Zone 3: The zone is windy, fair to partly cloudy, and seasonal.

Zone 4: Western and eastern parts of the zone see precipitation under variably cloudy skies, and temperatures are seasonal to above but cooler east.

Zone 5: The zone is variably cloudy and seasonal with precipitation west.

Zone 6: Fair skies accompany temperatures seasonal to above.

Zone 7: The zone is seasonal to above and fair under a high-pressure system.

Zone 8: Central Alaska is seasonal with scattered precipitation, eastern areas are windy and cold, and western parts of the zone are fair and seasonal. Central and eastern Hawaii see scattered showers, and temperatures are seasonal.

November 6–12, New Moon

Zone 1: The zone is mostly fair with temperatures seasonal to below and a chance for scattered precipitation.

Zone 2: The zone is variably cloudy with temperatures seasonal to below.

Zone 3: The zone is mostly fair to partly cloudy with precipitation west and temperatures seasonal to below.

Zone 4: Western skies are overcast and windy, central parts of the zone are fair to partly cloudy, and eastern areas see precipitation.

Zone 5: The zone is variably cloudy with precipitation east and temperatures seasonal to above.

Zone 6: Temperatures seasonal to below accompany fair and windy skies.

Zone 7: The zone is fair and windy with temperatures seasonal to below, but much warmer in the eastern desert, where there is a chance for precipitation.

Zone 8: Central Alaska sees precipitation, some abundant, and the rest of the state is fair to partly cloudy and seasonal. Western and central Hawaii are fair as temperatures rise, and eastern parts of the state see precipitation, some locally heavy.

November 13–20, 1st Quarter Moon

Zone 1: The zone is fair south, with northern areas seeing precipitation, some abundant; temperatures are seasonal.

Zone 2: The zone is mostly fair with temperatures seasonal to below and a chance for precipitation.

Zone 3: Temperatures are seasonal to below under fair skies.

Zone 4: Eastern areas have a chance for precipitation, and the zone is fair to partly cloudy with temperatures seasonal to below.

Zone 5: Variable cloudiness yields scattered precipitation, and temperatures are seasonal to below.

Zone 6: Temperatures are seasonal to below and skies are variably cloudy.

Zone 7: Eastern areas are cloudy with scattered precipitation, and western and central parts of the zone are mostly fair with a chance for precipitation; temperatures are seasonal to below.

Zone 8: Central Alaska is stormy, eastern parts of the state are windy, and western skies are fair; temperatures are seasonal. Central and eastern Hawaii are seasonal with showers, some locally heavy, and western parts of the state are fair.

November 21–27, Full Moon

Zone 1: Southern areas are mostly fair, and northern parts of the zone are cloudy with scattered precipitation; temperatures are seasonal.

Zone 2: The zone is partly cloudy south and central, fair north, and seasonal.

Zone 3: The zone is generally fair to partly cloudy and seasonal with a chance for precipitation west.

Zone 4: Variably cloudy skies accompany temperatures seasonal to above; western areas see scattered precipitation later in the week.

Zone 5: Western and central areas see scattered precipitation, and the zone is seasonal.

Zone 6: Temperatures are seasonal to above and skies are fair to partly cloudy with a chance for precipitation in central areas.

Zone 7: Southern coastal and central areas of the zone see precipitation, eastern areas are fair and windy, and the northern coast is windy with a chance for precipitation; temperatures are seasonal to above.

Zone 8: Central Alaska is stormy, and eastern and western parts of the zone are mostly fair and seasonal. Hawaii is mostly fair and seasonal, eastern areas see showers and thunderstorms.

November 28–December 4, 3rd Quarter Moon

Zone 1: The zone is seasonal, northern areas are windy, and southern areas see scattered precipitation.

Zone 2: Weather is seasonal under variably cloudy skies.

Zone 3: Eastern areas see scattered precipitation, and the zone is seasonal and fair to partly cloudy.

Zone 4: Precipitation centers northwest, eastern areas are cloudy with scattered precipitation, and temperatures range from seasonal to below.

Zone 5: Temperatures are seasonal to below and the zone is mostly fair with some cloudiness east.

Zone 6: Western skies are partly cloudy, while central parts of the zone are mostly fair; eastern areas are very windy with precipitation followed by cooler temperatures.

Zone 7: Northern coastal areas have a chance for showers, but western and central skies are mostly fair, while eastern areas see precipitation.

Zone 8: Alaska is windy and seasonal with fair to partly cloudy skies. Hawaii is fair to partly cloudy, windy, and seasonal.

December 5–12, New Moon

Zone 1: The zone is fair and seasonal.

Zone 2: Northern areas are fair, and central and southern areas see scattered precipitation; temperatures are seasonal.

Zone 3: Temperatures are seasonal to above and the zone is fair to partly cloudy.

Zone 4: Variable cloudiness prevails across the zone with a chance for precipitation west; temperatures are seasonal to below, but warmer east.

Zone 5: The zone is fair to partly cloudy and seasonal with a chance for precipitation west.

Zone 6: Temperatures range from seasonal to below under variably cloudy skies with precipitation west.

Zone 7: Northern coastal and eastern areas of the zone see precipitation, and skies are variably cloudy with temperatures seasonal to below.

Zone 8: Central Alaska is stormy, eastern areas are cloudy with precipitation, and western parts of the state are cold. Hawaii is cool and windy with precipitation.

December 13–20, 1st Quarter Moon

Zone 1: The zone is windy and seasonal with precipitation.

Zone 2: Northern areas see precipitation, and central and southern areas are fair to partly cloudy; temperatures are seasonal to below.

Zone 3: Temperatures are seasonal and the zone is windy under fair to partly cloudy skies.

Zone 4: Western areas are stormy, the Plains see scattered precipitation, eastern areas are partly cloudy, and temperatures are seasonal to below.

Zone 5: Much of the zone sees precipitation, some abundant, under very windy, cloudy skies, with temperatures seasonal to below.

Zone 6: The zone is variably cloudy with precipitation west and central, some abundant, and eastern areas are windy; temperatures range from seasonal to below.

Zone 7: Western and central areas see precipitation, and eastern areas are partly cloudy and warmer than the rest of the zone.

Zone 8: Central and western parts of Alaska are stormy, and eastern areas are mostly fair and windy. Hawaii is cool and cloudy with showers.

December 21–26, Full Moon

Zone 1: Temperatures are seasonal to below with fair skies north and precipitation south.

Zone 2: The zone is windy with precipitation and temperatures are seasonal to below.

Zone 3: Western and central areas have a chance for precipitation, and some eastern areas see abundant downfall; temperatures are seasonal to below.

Zone 4: Western parts of the zone are cloudy and windy with precipitation, the Plains are fair to partly cloudy, and eastern areas see scattered precipitation; temperatures are seasonal.

Zone 5: Eastern and western areas are fair to partly cloudy, and western parts of the zone are cloudy with precipitation; temperatures are seasonal.

Zone 6: Temperatures are seasonal to below, with stormy conditions east, mostly fair skies central, and windy with precipitation in the west.

Zone 7: The zone is fair to partly cloudy and windy east with temperatures seasonal to below; northern coastal areas see scattered precipitation.

Zone 8: Western Alaska sees precipitation, and central and eastern areas are variably cloudy; temperatures are seasonal. Much of Hawaii is windy with precipitation, but eastern areas are mostly fair; temperatures are seasonal.

December 27–31, 3rd Quarter Moon

Zone 1: Northern areas are cold with precipitation and high winds, and southern areas are more seasonal with precipitation.

Zone 2: Southern and central areas are very windy with precipitation, and northern parts of the zone are partly cloudy; temperatures are seasonal to below.

Zone 3: Western skies are partly cloudy, central areas are cloudy and windy with precipitation, and eastern parts of the zone are fair to partly cloudy; temperatures are seasonal to below.

Zone 4: The Plains are partly cloudy with precipitation, eastern areas are fair to partly cloudy, and western parts of the zone are cloudy with precipitation; temperatures are seasonal.

Zone 5: Western and central areas are variably cloudy, eastern skies are mostly fair, and temperatures are seasonal to below.

Zone 6: Eastern parts of the zone are windy with precipitation, and central and western areas are fair to partly cloudy; temperatures range from seasonal to below.

Zone 7: Central and eastern areas see precipitation, some abundant, and western parts of the zone are variably cloudy with a chance for precipitation.

Zone 8: Eastern and central Alaska are stormy, western areas are mostly fair, and temperatures are seasonal. Hawaii is very windy and cool with precipitation.

About the Author

Kris Brandt Riske's complete bio appears on page 183.

Economic Forecast for 2010

By Dorothy J. Kovach

You never miss your water, until your well runs dry.

~Wynona Carr

In 2010, batten down the hatches, because we are going to see the next floodgate breached. No matter how hard we try to stop the flood of economic woe, the problem cannot be fixed by adding more money. If we don't collectively take responsibility for where our money is going, the American dream will be over.

For many years, the masters of Wall Street have pulled the wool over all too many eyes. They led so many to believe that it was their mathematical genius that was keeping us well fed and rich. We believed they knew more than we did, but now we know the truth. And so many astrologers, including this one, predicted this carnage well in advance, because we know that the stars and planets above us influence our economy.

It is time to face the truth: we bet the farm, and now we are going to have to dig ourselves out of a debt that is as vast as the stars above. And to add insult to injury, we live in a world where the little guy pays and pays, while the big guys find loopholes.

For too many years, Americans have believed that a positive outlook or "magical thinking" could keep the markets up. We did nothing to protect our money; instead, we just handed it over to strangers and hoped for the best. Like the ancient emperor Nero, who chose to play his fiddle while Rome burned, we played our video games and pretended the "smart" guys with the Ivy League MBAs knew what they were doing.

Meanwhile, these captains of industry were collecting hefty stock packages as they shipped our factories and jobs abroad. In late 2008, when Congress hastily bailed out the very bankers that robbed us, we should have been incensed! Instead we passively allowed them to make this downturn longer than it needed to be by paying top dollar for assets that were, by the bankers' own admission, rotten to the core. Instead of making sure we got a fair deal for the taxes we pay to support these bailouts, we allowed our Congress to move us away from free markets to government-owned markets. By putting our tax dollars down a black hole called TARP (Troubled Asset Relief Program) and various bailouts, the odds are greater than ever that this present downturn will become long, protracted depression, much like the decade-long one the Japanese have just survived.

As long as we continue to drive around in borrowed cars and live beyond our means in borrowed houses, we are living on borrowed time. It is up to every man and woman who owns so much as a 401k to take responsibility. Know what is in your portfolio. Know the pitfalls of investing in markets. Know when to get in and get out. And above all, take control of your financial future. We can no longer look the other way and pretend others know better. They don't. As we saw in the past few years, it doesn't matter how good the company is; if 70 percent of stocks follow the trend of the general market, then it is important that each and every person know not only what the general market is doing, but everything they could possibly know about their own investments.

The forecast for the year ahead is not pretty. But I'd like you to keep one thing in mind as you read on: We have so much to be grateful for in this country. We have some problems, but it is our ability to get along—as diverse as we are—that helps us rise above the troubles and call ourselves united. The natural borders of the United States make us more secure than most countries. We share a common language, and above all, we must never forget that within our borders lie some of the most fertile lands and minds anywhere in the world. We are protected on two sides by great oceans. We are able to grow and feed the world. It is this bounty that will bring us back from these challenging times.

However, we won't get through this slump without hard work. For far too long, we have been led to believe that, with the proper advice, we could just park our money and it would thrive. This kind of pie-eyed optimism is typical of the thinking that permeates during bull markets. We believe we can do anything, that things are somehow different than they have ever been before. We believe there will never be a downward cycle, or if there is one, that we can somehow, with the right help or the right person or the right hedge fund, beat the system—if only that were true! Well, it's time to get real. The boom is over. We are now in preservation mode.

At the time of this writing in early 2009, we should be extremely careful before we put our money back in the markets. We are in the midst of a real bear cycle. Still, we can expect that, regardless of hard-to-come-by funding, there will be breakthroughs in the medical arena. Look for companies related to new forms of medicine to do well. We want to watch these companies closely, because we haven't hit the bottom yet. While medical companies should improve in the market, I would be very wary to invest until much later, at least until the end of 2010, when the dust starts to settle.

South Node in Cancer

The South Node shows where things "drain." Astrologers always keep a close eye on the South Node because wherever it is in the chart, and whatever corresponds to that area, will bring pain. In 2010, the South Node will continue its backward journey through the United States' monetary house. This placement brings with it the "empty pockets" syndrome, and it will make us nostalgic for the good old days of easy money.

With the South Node in Cancer, the sign of housing and real estate, we will continue to see the air drain out of the real estate bubble. The South Node is very important, because whatever it touches loses a sense of reality. When the South Node is with positive planets (such as Jupiter, Venus, or the Sun), we live in a world of illusion. The last time the South Node was with Jupiter, in 2005, authorities like Ben Bernanke informed us that there was no real estate bubble. We know now how wrong our top economists were! After all, the biggest investment Americans make is not in the stock market, but in their homes. Thus, once the South Node contacted the reality planet of Saturn in the fall of 2007, real estate took a flying leap off a cliff, pulling every other sector of the market down with it!

The South Node in Cancer throughout all of 2010 will do what bailouts cannot: bring down real estate values. This may not sound like such a good thing to those watching the equity in their homes disappear, but the fact remains that home values are too high. Loan mitigation can never be the answer. Housing must become affordable again. Only when this happens will our credit markets improve.

When young people cannot afford to buy and pay for a home in a reasonable amount of time, there is a problem. When home prices continue to rise, we eventually out-price a portion of the buying public. When the entry-level buyer is excluded from the real estate market by high prices, a lopsided situation develops. When housing becomes unreachable in affluent towns across the country, neighborhoods become devoid of the laughter of children and schools close because young families can't afford to live in high-price districts.

Our government should embrace lower-priced real estate instead of fighting it. Given the shrunken wages of the past two decades and the inflated real estate prices, we are sabotaging our young people from ever having real financial independence (unless they are lucky enough to be born rich). When the most viable segment of the population—first-time buyers—is unable to buy homes, the country's future is jeopardized by an unstable foundation. It is these high prices that the government wants to keep propped up.

Low interest rates are not the answer. Prices must come down. Even though the average mortgage rates are lower than back in the 1970s, payments adjusted for inflation are up 78 percent. That beach shack that went for $19,000 back in the day now goes for half a million. The sand states—Florida, California, Nevada, and Arizona—where prices rose the most in recent decades, will continue to see the biggest drops. The faster these prices recede, the quicker we will see a recovery.

Chinese Year of the Ox and Tiger

I have worked with the lunar calendar for my predictions for many years, so I thought I would share some with you this year. The start of the lunar calendar system is based upon the actual time that the Moon and Sun collide each winter. In temperate climates, the weather begins to get warm, and we often hear of a "false spring," which some believe is the new energy arriving. Just as there are twelve signs in the western zodiac, there are twelve signs in the Chinese zodiac: Dragon, Snake, Horse, Ram, Monkey, Rooster, Dog, Rat, Ox, Tiger, and Pig. Like the Western system, the Chinese system of astrology is a vast ocean, too deep for us to fully chart in this short space.

Since our purpose is to find ways to get ahead in these turbulent markets, though, it is wise to know when the lunar year begins and ends, because in the Chinese system, we want to emulate the qualities of whichever animal is in charge. As this annual comes out in the fall of 2009, we will be finishing up the second half of the year of the Ox. The qualities that are given to the Ox are the same as those given to the Cow. The Ox is noted for its constancy and practicality. It endures all troubles patiently and sweetly, and patience and fortitude will be rewarded.

To get ahead in the year of the Ox, we will do well to apply calm resignation to our activities. Holding back just a little bit and watching what we say in order to give encouragement will bring rewards. Those who go about their tasks in an even and steady manner, and above all, those who are not easily riled, will be rewarded. We would do well to follow the sage advice of Ben Franklin, and always remember that "a penny saved is a penny earned."

Even though the Ox is related to the bull, it is not to be confused with bullish behavior. Instead, the Ox brings what we call "sideways markets," where every bit of bullishness is followed by protectiveness and profit taking. Since the Ox is such a slow and steady animal, he gives rewards to those who take little steps. In markets, this implies increasing savings and decreasing risky speculation.

Since the Ox is a very large and timid animal, risk is rarely rewarded in such years, and the last quarter of 2009 is no exception. On the bright side, the steadiness of the Ox indicates that at least events will not be as radical as those we saw in 2008. It is this patience that we want to emulate in the second half of 2009 and the first month of 2010. In other words, hold back. Even if the market recovers a bit, it is not yet time to take risks. Above all, we know that while the Ox is in charge, we will not see the market reach bottom. We will be best served by sticking to only the safest of investments, so that we can best be prepared for the year of the Tiger.

On Valentine's Day 2010, the year of the Tiger begins. Those who know the Chinese zodiac will take a deep, bracing breath because they know it is difficult to predict what this breathtaking and dangerous creature will do. We all need to follow the Girl

Scout motto and be prepared. This will be a metal "yang" year. Like the poker burning in the fire, metal years can be red hot, but metal is not good for stocks.

In such a climate, we may see any recovery from the year of the Ox reversed. One thing is certain—we must be on our toes at all times to avoid being crushed during the year of the Tiger. It is no time to be complacent. It is a time of great risk. For those of you who are ready to stay silent and do their homework—just like the big cat watching the action from above before pouncing swiftly on its prey—there may be BIG rewards. A Tiger year is not for the faint of heart. It is a time when careful observation followed by swift action can bring great rewards; it is often the end of one era and the beginning of a new one.

However, during these times, many people can act without really thinking. It is no surprise that the plans for agreements, like the Munich Pact, are made and then torn to shreds in Tiger years. Walls are erected that separate people from each other. All too often, during a Tiger year, countries become "hungry" and march into others' territories. Hitler annexed Austria and Czechoslovakia in a Tiger year, as the leaders of Europe looked on passively and hoped that peace might be achieved by appeasing Hitler. Again, this is not the year for passivity but for quiet observation and action when needed.

In 2010, we should keep a close eye on China. The last metal yang year was in 1950, when UN forces (mostly comprised of American troops) got into a border dispute that some called a proxy war with China in Korea. With unemployment now rising due to the worldwide economic downturn, there will probably be much more unrest in China. Whether China will decide to stretch its borders at the expense of its neighbors, like Taiwan, is anybody's guess. But once again, this is no time to be complacent. There always seems to be a lot of saber-rattling in such years, so it's wise to keep an eye on the defense industries. War industries usually grow during Tiger years.

Such radical times can bring us to a solid bottom. It will be in our best financial interests to play it as safely as we can, because markets are places of trust, and there will be no faith in the markets' ability to rise substantially until the end of 2010. This tigress is metallic in composition. Metal years are noted for being very hard on equities. Therefore, we will be best served waiting to dipping our toes back into the stock market until the end of the year.

Jupiter and Saturn

Since the dawn of mankind, people with knowledge of the planets and their movements have watched with great apprehension when the two Titans—Jupiter and Saturn—met in the skies, knowing that a turning point was on the horizon. This is because they knew that when the two slowest of the visible planets met, whatever trend that was in progress, whatever weather was predominant, whatever crop was doing well—things were about to change. Unlike the financial leaders, astrologers knew that this change was indelible and that its functions would stick like glue for two whole decades (the time it takes Pluto to transit through a sign). And above all, astrologers knew that to fight such planetary magnitude would only prolong and worsen the inevitable slump.

Like it or not, everything in business is controlled by two ideas: (1) the principle of expansion, known as "bull markets," and (2) the principle of contraction, known as the deadly "bear market." Jupiter is our bull, while the ringed planet, Saturn, is the bear. Throughout 2010, Jupiter and Saturn are in a standoff, and they're both tied in a very challenging square to Pluto.

Unfortunately, when Jupiter met up with Saturn back in 2000, the markets had been going through the longest boom period in the history of the stock market. And that boom was directly influenced by Jupiter, the planet of luck and expansion. The Jupiter-Saturn meeting ended all that. Had our leaders only prepared themselves for the inevitable contraction (Saturn), the pain we're experiencing now would probably have been easier to manage. After all, we all know that the longer we look the other way when trouble is on the horizon, the bigger the problem gets.

Jupiter

Jupiter is the god of bull markets. He is the great expander, and whatever he touches brings great riches. We might even say that he has the "Midas touch." Whenever Jupiter is in charge, no matter what stock we buy, it goes up, and the market always seemed always to be going up. Under his influence, we think we can do anything. In recent years, everybody had money—even if only on paper—and we were willing to take big chances. Venture capital was loose, and workers were hard to come by. Jobs were plentiful, businesses looked abroad to meet the demand for products and services. Some organizations were forced to import people from abroad just to keep up with the pace. Since 70 percent of stocks move with the market, even lousy companies with terrible balance sheets profited. Everybody believed in his or her ability to become rich. In other words, when Jupiter is exerting influence, optimism is the watchword of the day.

All too often, though, the more money and power one has, the more he or she believes they are somehow above the laws of nature. When Jupiter, Saturn, and Pluto all danced through Libra (an air sign that favors intellectual property) in the early 1980s, a long boom began. Libra is an air sign, so the cause of the boom was due to intellectual advancement. When Jupiter met Saturn thrice in 1980, we saw innovations to the computer and, subsequently, the technology boom of the 1980s and '90s. This all ended in 2000. At this time, Jupiter once again met up with Saturn, but this time they met in the practical and down-to-

earth sign of Taurus. Technology stocks dropped like a rock. The bankers on Wall Street used everything in their power to create a boom. After 9/11, instead of telling us to save our money, American's were told to be patriotic and spend, spend, spend! Interest rates were left artificially low, while all sanctions with China were dropped. Soon, cheap imports flooded our stores.

Even though our wages had not kept pace with property values, we could always borrow against the equity in our homes. Anybody who could sign their name could qualify for a loan. Banks looked the other way when it came to collateral. Markets were propped up by fraud. It is time to face the facts. We have had far too much fun on borrowed time. It is as if we got into our daddy's liquor cabinet, and 2010 is the hangover. Unfortunately, again, ours was a Jupiter problem. We trusted when we should have been skeptical; we spent when we should have saved.

In 2010, Jupiter will be in his own sign of Pisces, a wet, soft, and sensitive sign. Pisces has a quiet kind of strength and a prevailing "last-minute" luck that may help to turn this nation around quicker than we expect. It favors silence and spirituality. It prefers to give rather than take. Pisces is the nocturnal home of Jupiter, and because it is wet, our feelings are often internalized. With Jupiter in Pisces, we are often brought to tears, we see the light, so to speak, and have a moment of breakthrough. This sort of personal reckoning may be needed before we can really turn the corner on these tough economic times.

Saturn

This is where Saturn comes in. Saturn is in charge of the law of return. Saturn is reality. Reality is hard to take, but it is a fact of life. Without Saturn, there is no success. It is as simple as that. In a nutshell, you work hard, you pay your dues, and then you get ahead.

If we could compare the two, Jupiter might be the guy who inherits the fortune. He never has to do a thing for it. He comes into the world with the silver spoon in his mouth. This guy will have all he wants—in fact, too much. This is the playboy son of the scions of industry, the gambler who, with a toss of the dice, throws away more money than some of us will ever see in our lifetimes.

Saturn, on the other hand, is the diligent worker who saves and works hard and makes a real mark in the world. He knows what it is to be poor. He is a responsible member of society, and he worked hard to get where he is. He doesn't try to shirk work by playing his way through life. When he does succeed, he knows where his success came from. He does not look to "beat the system" and knows that riches don't come overnight. They're built up by taking regular little steps and by making sacrifices. He also knows that riches can never be fully appreciated if they come without effort.

Saturn enters Libra on October 30, 2009. Even though Saturn is not known for its "drop in the lap" good fortune, believe it or not, the world will be very fortunate to have Saturn entering Libra, because Libra is all about fairness. This is the time when the little guy gets his day in court. However, justice does not come easy. Those who have been patiently waiting for their fair share finally get tired of waiting. Libra is not black and white, but shades of gray. This is why Libra is not just the sign of peace, but also the sign of war. When Saturn is in Libra (as it will be through October 2012), the word *conflict* is often on people's minds. It would be nice if we could bring about change through peaceful means, but revolutionary transformation is all too often bathed in blood. Hopefully, we listen to others' viewpoints and compromise.

Saturn is at its most powerful in Libra. It demands accountability. In terms of markets, we will probably see new regulations reigning in the so-called "free" markets. This may bring extra paperwork and red tape. Since we know that markets boom when Big Brother looks the other way, this does not bode well for a V-shaped, or quick, recovery.

Graft is rampant in bull markets. When Saturn is in Libra, justice and fairness are sought. Government actually looks out for the little guy by putting protective regulations in place. This is good news for the small investor. Unfortunately, markets sour when somebody is looking out for the little guy. Short sales are still ruling, because Saturn is so strong in Libra, thus we are all cautious. Markets are naturally more bearish, more careful, and more precision-oriented than they have been in thirty years. For long-term investors, this means it is not yet a buy and hold market, but instead a step-lively one, thus only for the most intrepid of traders.

Saturn and Pluto at Odds

Saturn is the ancient planet that rules the grave. Pluto is the modern planet of death. When these two are either conjoined or at odds, they have been known to bring extremities.

It will not be just homeowners defaulting on mortgages in 2010. We will witness entire cities going under. Cities derive their income from property taxes; when nobody can afford to pay their taxes, cities suffer. When states, counties, and cities have budget shortfalls, they can't afford to pay employees. That means schools, police, and fire departments are forced to cut staff. This inevitably leads to a corrosion of order. In such cases, cities become less user-friendly on the heels of default, be it by individuals, businesses, or municipalities. Saturn and Pluto are the harbingers of default. The infamous New York City bond default of 1975 took place during a hard aspect of Saturn and Pluto. When it becomes harder and harder to get ahead, people do what they have to do. With this in mind, we can expect crime to be on the rise.

As previously stated, Saturn is in charge of all contraction, and when Saturn is in strong aspect to anything (as it is with Pluto in 2010), that sector will either dwindle or become most difficult to obtain. Pluto has to do with monopolies, big business, the mining industry, and those things we loathe, like debt and taxes. When Pluto is prominent, as it is in 2010, it brings great shifts in the marketplace. When things are going too well, Saturn can be like a lead balloon, bursting our optimism, but Saturn's steady force also acts as the string that keeps the balloon from floating away, keeping us grounded.

Indeed, Saturn is all dark and somber, but because we know what to expect, we can at least count on him to give us structure. Pluto, on the other hand, is here to just plain kill us. Even though the astronomers have decided to remove Pluto from the pantheon of the solar system, wise astrologers know not to ignore his glances, because his looks can and do kill. When Saturn and Pluto are at odds, take my word for it, there will be no "Skip to ma Lou, my darling." With these two in hard aspect, those who have refused to face the reality of the economic climate will get more than a hard kick in the pants.

Every time these two meet, the economic repercussions are enormous. I suspect that when they cross, they will do their duty and shake out the last holdouts of the bulls along with their magical-thinking counterparts. You know the type: they're the dinosaurs who still "believe" in the market and who thought that markets could still be manipulated. They're the financial cheerleaders seen on some of the television channels— some of the same people have been spouting of a "soft" landing since 2007.

We need to make something real and tangible to sell—real products, not complicated debt instruments or financial schemes. The idea that we could just spend money without having to sell anything helped get us into this mess in the first place. America must work to replenish its industrial base. We need to have a real balance of imports and exports. We cannot thrive in this country if ships continue to leave our harbors empty.

It is as if our protection is violently stripped away, and we are left standing naked as a gale force wind blows us down. It is no wonder that the last time Pluto and Saturn were in hard aspect, we watched in stunned horror as the Twin Towers of New York's World Trade Center came tumbling down in a nightmarish terrorist attack. In the financial world, we witnessed an economic tsunami. That crisis did so much damage that the markets were forced shut, only to drain upon reopening.

Another thing that the Saturn/Pluto cycle seems to enforce over and over again is that we cannot depend upon the goodwill of others. Good intentions are not enough. When Saturn is in square to Pluto, we are forced to admit how very vulnerable we all are. Is it any coincidence that China has been increasing its military spending by leaps and bounds? Sabers are always rattled during Saturn/Pluto occasions, so don't expect defense industries to totally shut down, even during a Democratic administration.

The Financial Quarters

The Ingress: Fall Equinox to Winter Solstice 2009

We could say it was the best of seasons, and it was the worst of seasons. In some ways it may be a bit of both. America's transition from being the great power in the world to being the biggest debtor nation didn't take place overnight. Our debts have been multiplying for more years than we can remember.

As the last quarter of 2009 comes to a close, the American public is worried. With so many people overloaded with debts that they may never be able to pay and others not making payments on declining assets, there is a huge problem. Essentially, we can work and work all we want, but until our government gets *their* house back in fiscal order, it will be pretty hard for people to make ends meet, much less get ahead.

Eventually, the government must realize that we have to increase exports to get out of the hole we are in. If the attitude in government is to ship jobs away and look the other way while the banks tempt the populace with loans they can never repay, it is only a matter of time before we have to pay the piper.

Some people think the United States is too big, too established to fail. The problem with being too big to fail is that when the biggest game in town does go down, they tend to take everybody else down with them. This can make a lot of those everybody elses hoppin' mad. At the opening of the fall season, there may be some threatening talk from our open enemies. At this time, Mars is in Cancer, a place where anger is turned inward and left to stew. This is really lucky for us Americans, because our enemies may want to harm us, but they will also be weakened by the downturn. This is a Darwinian market where only the strongest will survive.

First Quarter: Winter Solstice 2009 to Spring Equinox 2010

Instead of a white Christmas, this one may be pink due to the many pink slips across the board in all industries. Every piece of good news will bring with it two pieces of bad news these days. Housing is still on the rocks. It may not be the greatest time to be spending money we don't have, even if it is Christmas. Basically, we are coming to the second Christmas in a row where it may be finally "in" *not* to spend.

Simply put, if you don't have it, don't spend it. However, we need to remember that it is always darkest before the dawn. Breaking the habit of spending may be easier said than done, but those who try will find themselves in far better shape. We are best served by being frugal this winter season.

Gold and its sister, oil, may be hitting some lows around now. Gold does not do well, historically, in deflationary times. With so many good dollars thrown after bad, let's hope hyperinflation is not in our future, either.

Second Quarter: Spring Equinox to Summer Solstice

If you can keep your head when the rest of the world loses theirs, you will get ahead in this cycle. We must never lose sight of what really matters: our loved ones and friends. Material items are not alive. While they are nice distractions, they do not bring us love.

It was almost ten years ago, in the days and weeks following 9/11, that we were told that the best thing to help the economy out of this crisis was to spend, spend, spend! It was just about a year ago, this February, that the labor department reported that those seeking unemployment benefits jumped to a record 4.9 billion. Many of those people have now used up their unemployment benefits, and those once-hardworking souls will be out on the street. If we are to get through this downturn intact, it may be time to get back to the basics—our family, our friends, our loved ones, and our community. If you are reading this in the fall of 2009, here's the bottom line: if you don't have it, don't spend it. We will do well to practice more thrift and less envy.

Digging our way out of our collective debt starts at home. No matter what the pundits tell you, we still have not yet reached bottom. The bottom will not come until our housing prices come down to more realistic levels. Our property values will probably continue to decline this quarter. Alas, this can spell real trouble to those who are already deeply in debt.

On the world and market front, it is easy to make mountains out of molehills, so despite the best of intentions, we may see some real hostilities mounting. This implies that the military and defense contractors may not be losing as much money as they thought given the stated pacifism of the current administration.

Third Quarter: Summer Solstice to Fall Equinox

Uncertainty is manifold. When markets are uncertain, they tend to go down. Ptolemy said that love and hatred make cause for mistakes. Sadly, it is going to take a lot longer than one season to get us out of the mess.

Americans need to prepare themselves for another year of fewer jobs to go around while the trickle-down economy sheds jobs and assets. Most wonder when this nightmare will end, as they watch both their fixed and liquid assets continue to drop in value. We cannot throw money at a problem that was caused by too much debt in the first place. The more money that is given to the irresponsible, the longer the downturn will last. So-called "bailouts" propped up many businesses that would have surely failed if these were truly free markets. Just remember this: unemployment always lags behind the market. Wise investors never forget that it is always darkest just before the dawn. With all the pessimism rampant, there may be some real bargains out there at the end of this year. Remember, we must be contrarians. Listen carefully to the news around you. The more pessimistic the news, the more negative the basic overall outlook on markets, then it is time to do our homework and start investing again—at the end of this cycle, but not until then. Happy Trading!!!

About the Author

Dorothy J. Kovach is a traditional astrologer who utilizes both Eastern and Western methods in her work. Best known for using the traditional methods of William Lilly, she called the market downturn in
2000 to the month, five years in advance. More recently, she warned
her readers to get off the markets by the beginning of 2008. She also
called the drop in both oil and the fall of the Euro, to a rather skeptical audience, back in April of 2008. If you would like to have a more
specific idea of what the markets are going to bring, sign up for her
newsletter, at www.worldastrology.net or write her at Dorothy@
worldastrology.net

New and Full Moon Forecasts

By Sally Cragin

At various times, I've taught a class in understanding lunar phases. Most people are aware of the Moon when it's full, or nearly full, but they're not so certain of the stages in between. That's the tricky stuff, but when you get a handle on those other weeks, you'll be much more efficient at planning your monthly activities. At this point, the class usually looks hopeful and cautious. And so I tell them: Let's keep it simple. Let's divide the lunar orbit into its two main components. When it's waxing, it's getting bigger. When it's waning, it's getting smaller. Okay, is the Moon outside getting bigger or smaller? (Here's where I usually get some correct answers—and some grateful nods.)

Ebb and flo, up and down, in and out. The rhythms of the Moon are the swiftest astrological transit at our disposal. Keep this simple and understand two basic principles: When the Moon is waxing, it's now time to build. When the Moon is waning, it's now time to withdraw. As soon as you understand that crucial pattern, you make the biggest leap in your education about the Moon.

I've spent my life staring at the Moon. In part, this is a benefit (!) of being nocturnal by nature. In the 1960s, when I was a child, it was the most exciting period of extraterrestrial exploration. I had an enormous advantage: when the Apollo missions were being launched, my uncle, Dr. James Trainor, was directly involved with this exciting branch of science. He was a space physicist for NASA, and he built things that were put on the various rockets shot at the Moon and beyond. If you've seen the movie *Apollo 13*, there's a dramatic scene set at Mission Control. That famously underplayed distress signal: "Houston, we have a problem" is immediately followed by a bustle in the control room. I recognized those guys when I saw the film—all of them were my uncle in the 1960s! Military-trim haircut, horn-rimmed specs, and a short-sleeved white shirt. Intense, but always in control. For one glorious interlude in modern times, the scientists and the astrologers were in accord about their fascination with the Moon.

As we charge through the third millennium, it's difficult to remember just how important the Moon was for this country in the 1960s and early 1970s. Getting there meant we would triumph—but over what? For thousands of years, human beings had stared at this bleakly mesmerizing celestial marker, the opposite of the Sun, its mirror image and night-sky twin. Cool versus hot, near versus far, pockmarked versus smooth. The Sun controlled the crops, but the Moon controlled the tides, the spawning fish, and even our own bodies. Our word menstrual derives from the word for "month," which derives from the word for "Moon," whose orbit equals a month.

So no matter how modernized we get, chances are our own (female) bodies will mimic that rhythmic twenty-eight-day pattern. And to get back to my class: here's the easy way to tell if the Moon is waxing or waning. If the Moon is waning, the shadow of the Earth is on the right side, and it gets bigger with each passing day. By the fourth quarter, the Moon looks like a lopsided "C." If the Moon is waxing, the shadow diminishes from right to left, and the earliest crescent Moon is like a "D" that gradually grows larger.

Full Moons and New Moons

> "There are certain times to record," musician Neil Young would tell writer John Rockwell in 1977. "For the longest time I only recorded on a Full Moon, and it always had the same intensity . . . . Everybody would get crazy."
>
> —Jimmy McDonough, *Shakey*

Full Moon madness is no joke, and Scorpio Neil Young was a canny customer to take advantage of the lunar rhythms with his band Crazyhorse. It takes someone with a strong stomach (or who was born on or around the Full Moon) to be totally in command when the Moon is full. As you are reading these forecasts, bear in mind that when the Moon is full in your Sun sign, others could have expectations of you that are out of line with what you think you can produce. For example, if you're a Sun sign Capricorn and the Moon is full in Capricorn in July, others may think you're feeling more practical, deliberate, and structure-minded than you are. On the other hand, during the time that Moon is full in Capricorn, you may be able to encourage the "Capricorn" traits in others. For example, encouraging someone to be more independent, more capable of taking care of themselves, or thinking more practically.

New Moons are entirely different. I knew a licensed practical nurse stationed at an Alzheimers' ward in a nursing home who made very acute observations about her charges' behavior and the lunar phase. She always said that during the Full Moon, the predictable activities occurred: increased volume, outbursts, trays spilling, hot tempers. But during the New Moon, the quietest patients got agitated and acted up. This always took the staff by surprise.

Once you've spent some time observing the world (and your own behavior) during a Full Moon, you'll be more attuned to the variations that occur during that pivotal New Moon time. The New Moon is also useful to monitor who comes into your life, what they want, and why they are needing help or encouragement or your talents. People can be tentative during the New Moon, but you can gain a lot of usable knowledge about what your activities and preoccupations will be during the next two weeks.

Lunar Phases

> The Moon grows from a sickle to an arc lamp, and comes later and later until she is lost in the light as other things are lost in the darkness.
>
> —George Bernard Shaw, *Heartbreak House*

If you are new to this topic, I'd suggest you copy some of these pages and cut out the highlights; then tape the pages onto your calendar to remind you of the quality of different phases and signs as they affect the Moon.

In a **waxing Moon** (from New Moon to second quarter; black on the left, white on the right, increasing amount of white on the right), think in terms of: making, doing, gathering, preparing, building, being hopeful, creating, buying, doing. In a drama, this part of the play is the first act or the "rising action." Simple farming lore: waxing is planting and growing; waning is weeding and harvesting.

Second quarter Moon can be a turning point in a project. Second quarter Moons are also a time when a new personality could interject themselves into your "well-laid plans," or an unexpected change in schedule or expectations could make you improvise a solution. Versatility is a key asset during this phase, although the day itself might be neutral. Watch what happens on either side of this particular day.

Gibbous Moon is the period between the second quarter and the Full Moon. The Moon looks like a lopsided peach. This is the most intense time, especially three and fewer days before the Full Moon. Projects and personalities get more forceful or ardent. This is not the best time for people to hear things. It may look as though people are listening, but the end result is . . .what? I'm sorry, my mind drifted there. Yes, focus is a problem. Don't assume people know what you're thinking or feeling just because you're thinking or feeling it so intensely!

The **Full Moon** means the Sun and Moon are aligned with Earth. The tide is high emotionally. The Full Moon is a great time for a party or a gathering of very different people. When it occurs on Hallowe'en weekend, prepare to buy much more candy than usual! This can also be a time to sell something and get the highest possible price; conversely, you could also be rooked by paying more than necessary for something. One of the great benefits of the Internet is being able to price-compare, and though there's nothing more tangible than seeing something right in front of you, you may find it online for less. The Full Moon is also a time for the "impulse" purchase. There are drawbacks to this Moon and exhaustion can be a byproduct. Don't forget, the word lunatic derives from luna.

Waning Moon phases signify completion, finishing, letting go, cleaning out, removal, diminishing, excising, discarding. Great activities for the fourth quarter Moon to the New Moon include decluttering your house, wrapping up unused items for charity, and going through piles of things you might have in duplicate (files, storage containers in the kitchen, clothes children have outgrown, workbooks, manuals to computer programs no longer used) is a very useful and proactive task.

The **fourth quarter Moon** is about getting results. If you find that you're less interested in pursuing a line of thought or action, those apathetic signals could be very strong during this time of the month. And if you find that something you've invested in—a relationship, a gamble, a change of direction—is losing traction or that those who are participating are less present, well, that's the waning Moon for you. There's no forcing a situation during this phase.

Balsamic Moon is the shadowy twin of the gibbous Moon. But instead of the Full Moon coming at us full tilt, the New Moon is looming. Feeling like you need to scramble or hustle could be a symptom. This is a great time to look for bargains and underval-ued commodities or objects. This is also a good time to bargain. I try to time my automatic deposits when the Moon is close to new, just to get more from my money.

The **"Dark of the Moon"** is the very last day (and sometimes the day before, depending on the time the New Moon actually occurs) before the New Moon. It's an accident-prone time, but it can be extremely useful for picking apart your subconscious and seeing what desires/themes reside therein. This is a time when you may want to bail on a project, which is fine. But my advice would be to wait until the New Moon so you won't have to reverse your position again. I also find this is a time when I feel at a lower ebb energy-wise, and my friends report the same symptom. Feeling less enthusiastic about a cherished activity is also a feature of this Moon. "La dee dah" ambivalence turns into the blahs.

New Moon is the ending and the beginning all at the same time. You may think this is a neutral time, but it's not, and the smart student of the Moon will make a point of taking notes during this lunar phase. How do you define endings and beginnings simultaneously? A job ends—a new job begins. A move. A journey someplace new. Starting a new class. This is also a time that is said to be great for meditation . . . once you can tune out the feelings of dread that occasionally accompany this phase! If you haven't cleaned your place or emptied a closet in the previous week-plus, now is finally a good time to go for it. This is also an emotionally draining time; if you are able, take things easy.

The Elements in Play with the Moon

When the Moon is in a fire sign (Aries, Leo, Sagittarius), you'll have energy for new projects (even during a waning phase), new people, and lots of activity. Go slow if you find you lose your temper about trivial stuff. Excitement comes easily, and impetuous behavior is the norm during these times. If you find that you're regularly "talked into" something during this lunar phase, be aware that you should probably get into the habit of saying, "I need to think about that, let me get back to you . . ."—and in your head, complete the sentence with, ". . . when the Moon is in an earth sign and I can be more practical!" In short, you can easily spend too much time on too much talk. Wasted energy is the downside of the Moon in fire signs.

When the Moon is in an earth sign (Taurus, Virgo, Capricorn), it's all about "show me the money." Activities that relate to personal security, banking, work compensation, health insurance, home insurance, and structures as relating to financial matters flow more easily during these times. Every time I sit down to tally up receipts or check on my check register, the Moon is in Capricorn. I find that particular Moon increases my tolerance for practical tedium and activities I'd otherwise find boring. This is also a useful period after the preceding hectic days of the fire-sign Moon.

When the Moon is in an air sign (Gemini, Libra, or Aquarius), social activities come easily. Even the shyest folks can reach out to others without flinching. This is a great sign for communication, talking, writing, and so on. It's also a good time to plan activities that involve a lot of other people. Following the earth sign Moon, this lunar phase can pep you up and give you renewed enthusiasm for a project that may have stalled or slowed down. However, watch out for frivolity, i.e., doing something twice quickly, rather than once slowly.

When the Moon is in a water sign (Cancer, Scorpio, Pisces), you'll want to wallow in feelings. My observations reveal that defensiveness (a trademark of the crab) is more likely when the Moon is in Cancer, but you may find that Moon in Scorpio is when people really react with sharpness. Still, this is a dynamic time for all kinds of creativity or appreciating artistic disciplines. The muses love this phase. The water sign difficulties can come with unintentional moodiness, sogginess, self-pity, and self-indulgence.

A few words about the void-of-course Moon. If you have also invested in *Llewellyn's Astrological Calendar*, you'll see a notation after the Moon symbol: v/c. That means the Moon is void-of-course and is not (for minutes or hours) making any major angles to other planets. This time is considered a freeform period and even well-made plans can easily dissolve like tissue paper in the rain. However, a void-of-course Moon is a great period for entertaining unusual ideas or crazy options or having an artistic breakthrough.

Eclipses!

Yes, eclipses can make people do very strange things—and if you were reading this two thousand years ago (on a Sumerian clay tablet, say), you'd still be nodding your head in agreement. Eclipses are another rhythmic event in the celestial choreography and as fascinating as they are, they still, to use a technical expression, freak people out.

Eclipses are traditionally associated with power transitions, unexpected collapses of government or organizations, and self-destructive behavior. In the year 1936, which in the United Kingdom is also referred to as the "year of three kings," there were five eclipses (the sixth was on Christmas Day of the previous year), which is more than the usual number. Eclipses can be associated with destabilizing activities and a wavering from commitment. In 2010, there will be four eclipses: January 16, June 26, July 11, and December 21.

. . . .

In conclusion, my advice is for you to keep a tab on these pages so you can refer to the comments as the New Moons and Full Moons occur. Don't be surprised if you're not "bothered" by one Full Moon or another, but keep track of your own personal responses. I often hear from clients or students that they are surprised at how their expectations for a Full Moon are out of line with their own direct experience. Full Moons and New Moons are useful, and these tips are the equivalent of a "tourist brochure" as you navigate your way through the night sky!

New and Full Moon Cycles for 2010

Full Moon in Cancer (with eclipse), December 31, 2009
The old year goes out with a bang and an eclipse, and this is the Winter Moon, also known as the Wolf Moon. It's a lunation that could prompt feelings of financial urgency, particularly among Cancer, Libra, and Aries individuals. And with Mercury retrograding (as it will through the next New Moon), you may seem more committed than you really are.

New Moon in Capricorn, January 15
Despite two retrogrades (Mercury and Mars), this Moon is excellent for starting a project relating to your domicile or your

long-range career plans. If you need more training in an area to become technically proficient, that will be very clear during this Moon (especially for Taurus, Capricorn, and Virgo, who will be craving opportunities to advance all year).

Full Moon in Leo, January 30

The year begins with what the Tewa tribe called the "Moon When The Coyotes are Frightened." You, of course, have nothing to fear as this Moon occurs in social, party-loving Leo. This is an excellent weekend to throw an impromptu "winter's nearly over" party. Aries and Sagittarius could be in a "joke-playing" mood, and if you want to know where the most fun is happening, speak to an air-sign person.

New Moon in Aquarius, February 14

Valentine's Day can be a fraught time, but with this New Moon in Aquarius, set your sights on unusual or eccentric experiences or individuals. You'll be able to see the merit in the quirkiest personalities. Your appetite for imaginative or creative experiences will be enormous, particularly for Pisces, Aquarius, and Gemini folks. Don't settle for the "conventional" answer, and keep in touch with faraway friends.

Full Moon in Virgo, February 28

Ordinarily, this would be a fine time to get your vehicle looked at, but with so much tension from Uranus (the planet of surprises!), you may want to keep your activities predictable. This is an excellent time to visit a new health center or to get good advice on nutrition, particularly if you have digestion issues. If you're an earth sign, Cancer, or Scorpio, schedule dental work for this Full Moon. Others may tell it like it is, and if you're a tender type or one who tends to read into things, you could be feeling sensitive. Virgo can be a workaholic type of sign, so if you've put off a complicated task, burn the midnight oil.

New Moon in Pisces, March 15

Pisces loves to help green things—or just things in general—grow. Feel like getting a haircut? Need a new pair of spring shoes or brightly-colored stockings? This New Moon is an excellent lunar phase for having a deeper-than-usual psychological insight about a friend or family member. There are a lot of folklore recommendations for this particularly phase (see August 24 for more), including brewing beer! However, some signs (such as Virgo and Gemini) might be in a mood to brew trouble during this Moon.

Full Moon in Libra, March 30

Time for a spring makeover, although Cancer and Capricorn people might want to wait until Wednesday or Thursday. This is the Planter's Moon, also known as the Egg Moon, and it's a helpful lunar phase for all kinds of surface treatments. Appearances count, and Libra's diplomatic nature could bring you some helpful contacts that, with gentle nurturing, could evolve into new opportunities. The downside of this Moon is terminal indecisiveness, which could afflict Aquarius and Gemini.

New Moon in Aries, April 14

If your appetite is a little off, your body may be telling you that it's time to lose the winter weight. Aries Moons bring out the impetuous side of everyone, and with a helpful aspect to Mars, feelings of ambition could be a saving grace for Aries, Sagittarius, and Leo. Quick decisions come easily, and if this decision is something temporary (haircut, trying a new computer program), definitely go for it. Impulse purchases are likely this New Moon, as are short-term investments.

Full Moon in Scorpio, April 28

This year, the Moon and Mars are at odds during this Moon, which the Algonquin tribe calls the Milk Moon. Aquarius and Taurus feel easily irritated during this phase, and it's an excellent time for activities related to shopping, healing, or breaking down a plan to its smallest components. If you are a list-maker, you'll be able to construct action-plans that include the order in which steps should be taken. Even more happily, this particular water sign Moon is very positive for emotional and physical connections. You may feel like dressing provocatively or find yourself winking at a stranger.

New Moon in Taurus, May 14

Spring Moons have an emphasis on appearance, and this Taurus Moon is super for making a long-term investment in appearance or some aesthetic component of your life. Even if you're not at all materialistic, this Moon can prompt a feeling of covetousness. Cancer, Taurus, Pisces, Virgo, and Capricorn all have unerring taste right now, and if you need to upgrade an expensive item, go for it.

Full Moon in Sagittarius, May 27

If you haven't planned a summer vacation, now's the time. The Colonials called this the Honey Moon or the Strawberry Moon, which is your signal to take a break. If you haven't gotten lost while wandering or explored a new culture recently, do so now. This lunar phase is all about travel, and with Mars in a compatible fire sign (Leo), "making a move" will be irresistible. This is also a great Moon for starting a diet. The risks that come to this Moon can be as simple as basic clumsiness. Are you in a rush? Do you need to be? Slow down.

New Moon in Gemini, June 12

New Moons signify fresh starts, and Gemini is all about conversation, interaction, and communication. This is an excellent time for getting in touch with a large group of people or making sure messages are coming across clearly. Those of you in the law or judicial profession will find that thinking clearly and being able to see both sides will come easily. This New Moon is another good period for stopping a habit, particularly one connected with your mouth or lungs (junk food? smoking?).

Full Moon in Capricorn, June 26

Summer is here, so why do you feel like working? Earth-sign Moons are excellent for labor, particularly tasks relating to systems, structures, and patterns. This was the Buck Moon for the Algonquin tribe, and if any sign can "buck" the system, it's Capricorn. Capricorn also rules the knees, so you may be tempted to take a long walk over uneven ground. This Moon can bring blunt talk, so if you're an Aries, Libra, or Cancer, your emotions could be raw. For Capricorn, Virgo, and Taurus, this is the perfect Moon for achieving (what you had thought was) the impossible.

New Moon in Cancer, July 11

A super Moon for domestic activities, baking and cooking, or making your home cozier. Like the other water-sign Moons, this is a fine Moon for brewing . . . beer, that is, although Cancer's insightful nature could prompt some signs (Aries, Virgo, and Taurus) to suspect brewing plots! Cancer and Scorpio are especially sensitive to others' needs during this lunation. This is an excellent time to begin a home-improvement project or purchase new ceramic dishware.

Full Moon in Aquarius, July 26

The Dog's Day Moon actually precedes the appearance of Sirius (the dog star) in the sky, but there will be plenty of opportunity to let loose and howl, especially for Libra, Gemini, and Aquarius. The wildest ideas fly, the happiest fantasies delight, and the unexpected guest turns up for dinner or with a need to bend your ear. This is a restless time for some, particularly Leo. Items to purchase include: electronic and/or labor-saving equipment.

New Moon in Leo, August 10

With a fresh start for fire and air signs and Venus and Mars united in Libra, this Moon could bring unexpected but harmonious partnerships. If you've put off having your vehicle serviced, now's the time. If you're tired of your same old warm-weather clothing, can you dress up your wardrobe by changing a hemline or adding some trim? Leo is in a great position to see all sides of a situation, particularly when dealing with their special Capricorn friends.

Full Moon in Pisces, August 24

The truth will out . . . in a roundabout way. This is an excellent Moon for making something artistic or to indulge in escapist literature, films, or music. Pisces Moons can prompt some signs (Gemini, Virgo, and Sagittarius) to feel raw, but Scorpio, Taurus, Pisces, Cancer, and Capricorn can be uncommonly perceptive and wise. It's also an excellent time for shoe-shopping—Pisces rules the feet—or getting an impromptu pedicure.

New Moon in Virgo, September 8

What happened to the summer? With September here, you still have an opportunity to get to those projects that have been on your list for the past few months. Doing things efficiently is a Virgo trait, and Capricorn, Scorpio, Cancer, and Taurus will be cruising with this Moon. This is another good time to think about stopping a habit, particularly one that has an effect on your health. Virgo: The two years of occupational/personal transition you weathered from September 2007 through July 2010 are in the past. Now you get to build on what you learned.

Full Moon in Aries, September 23

Shine on, Harvest Moon, and burn the midnight oil, Aries, Leo, and Sagittarius. This Moon will bring new perspective on old projects. It's an excellent time to take on a short-term project or a task that can be turned around in a day or so. Libra should be cautious about "misspeaking" or signing up for activities that you're less than passionate about. Aries: You're in "command mode." Make sure the troops understand the task before you set 'em loose.

New Moon in Libra, October 7

Libra is everywhere! Sun, Moon, Mercury, Venus, Mars, and Saturn. Finding a method of collaboration or a "third way" should come easily to many, particularly air signs and Sagittarius and Leo. A desire for new clothes or more elegant, comfortable, and dignified attire could be an amusing distraction. Starting from scratch is better than building on a half-finished idea. Capricorn, Aries, and Cancer: You may be a little sensitive this month, which is amplified during this lunation. Don't hesitate to ask for clarification if you feel you're not getting all the information you need.

Full Moon in Taurus, October 23

The Hunter's Moon cranks up our materialistic impulses. Last month's New Moon in Libra awakened an appetite for an improved wardrobe, and this Moon will have the same effect on your desire to improve your environment. Practical matters take precedence, however, so get your vehicle attended to, look at your credit rating, and only make purchases if they're "investments." (Yes, this can include a haircut with the most talented employee of the salon.)

New Moon in Scorpio, November 6

Another excellent Moon for brewing, according to the folklore, or cutting back on expenses. Venus and the Sun and Moon are in synch for this New Moon, so you could be the recipient of confidences. This Moon can also bring out your practical side, so if there's clutter, deadwood, or outworn objects in your home environment, it will be easy to discard. Leo and Aquarius may need to take a stand with someone who hasn't been "getting" them.

Full Moon in Gemini, November 21

This is a restless Moon, and those with a lot of air-sign influences in their chart may find it tempting to be indiscreet about someone else's situation (Virgo and Pisces are most at risk). However, the positive tasks you can accomplish this month include stopping a habit or buying a machine that helps you communicate better or take short journeys. If you commute on the bus or train, does it make sense to get the monthly pass? Gemini: you've got some choices to make this month, but you won't yet have all the information you need—ask some more questions.

New Moon in Sagittarius, December 5

During November, as Mars moved through Sagittarius, the fire signs had lots of energy for new activities. This lunation brings excellent potential for starting a project, planning a long trip, or exploring the possibility of further education. If you know someone from another cultural background, socialize! (Especially if they can cook!) All kinds of physical activity are indicated, especially those that favor the upper thigh (horseback riding, skiing, bike riding, hiking). Sagittarius should use this interlude to plan a range of activities you would like to do in the next three months (bearing in mind how quickly you can get bored).

Full Moon in Cancer, December 21

The solstice and the Full Moon are excellent astrological ingredients for family gatherings. You'll be able to listen carefully to loved ones, particularly those of you with a lot of water in your charts. The simplest pleasures are best during this Moon phase and sign. Making cookies with a kid will provide more fun than wrestling with pastry dough all on your lonesome. Ancient folklore has many suggestions for this lunar phase including making sauerkraut, canning or preserving, or brewing beer. If you're not handy in the kitchen, you may find your senses and appetite sharpened for consuming those items.

About the Author

Sally Cragin, as Symboline Dai, writes "Moon Signs" for the Boston Phoenix *newspaper chain. She is available for private consultations and can be reached at Moonsigns.net/*

2010
Moon Sign Book
Articles

The Garden: A Base of Spirit and Sustainability

By Penny Kelly

At the age of six, I trailed along behind my mother and her shovel, lugging two buckets that weighed almost as much as I did. One had seed potatoes in it, the other was half full of a white, granular material she called "fertilizer." She would dig a hole, and my job was to drop in a potato, then a handful of fertilizer.

She worked her way down the row and when we reached the end, we worked our way back, filling in each hole and tamping the soil snugly over each potato. When the potato plants came up, I watched her dust the growing plants with rotenone to keep the potato bugs away and pull soil up around the base of the plants in a process I later learned was called "hilling."

Quite some time after this, once again out in the garden, I noticed with dismay that the potato plants were dying without producing a single potato. My disappointment was keen. The bean seeds had produced beans, the tomato plants were filled with green tomatoes (but at least they were growing), and although the carrots and beets were miniatures of what I was familiar with, they were there and getting bigger. Only the potatoes had failed.

My mother seemed to ignore the potatoes and never said much about them while my father faithfully continued to weed and till alongside the row where they had been.

One Saturday afternoon in early October, my mother took a giant fork out of the garage. She told me to get a bushel basket and come to the garden. When I arrived she was turning over the row where we had planted the potatoes, and I was stunned to discover the ground under the dried, shriveled remains of the plants was loaded with fresh, new, tender-skinned potatoes. With each thrust of the potato fork, more crisp, golden potatoes emerged from their hiding place in the "hill." It was like digging for gold! For a little while, my mother and I immersed ourselves in what became a game, turning over the soil, sifting through it excitedly.

I was hooked right there. Although the garden lost its glamour in my teen years, its deep pull resurfaced in the depths of Detroit some years later.

In the asphalt jungles of the city, life either hardens and conforms to artificiality, or it sets up a great echoing call for something unnameable. Like the internal sonar of some blind creature, I found my soul reverberating, bouncing off the cement walls and streets in confusion.

A few pots of tomatoes on my second-story deck were supposed to bring relief from the monotony of cement, asphalt, and days in the office. In the midst of my frustrated attempts to enjoy the process and keep them alive long enough to reap a small harvest, I found myself suffering the powerful understanding that the deck was not a natural or normal environment for giant tomatoes in tiny pots nor for country girls transplanted to the city. I could *feel* the agony of those tomatoes. They did not even come close to the healthy, vibrant plants that had grown in the gardens of my youth. Every day, the plants needed to be revived and, in spite of my constant watering, it was not enough. I went away on a hot summer weekend; when I returned, they looked like they had been fried. I grieved as though I had lost my best friend.

A couple weeks later, I was hit by a car while riding my bike. That was the end of my life in the city. I moved home to northern Michigan and spent three months recovering. It was the beginning of my wisdom.

Slowly, I came to realize that something inside was calling me to the land, the soil, and plants. I wanted to write, and at first it seemed that giving in to the call of the land was a huge distraction from any possible writing career. However, it was only when I gave in and started to plan and save for a house in the country where I could garden all summer and write all winter that life began coming together for me. We bought a small farm with a couple of vineyards, and I eventually gave in to the call of those vines, which led to my first book, *The Elves of Lily Hill Farm.*

The more time I spent in the vineyards and gardens, the more I began to see and understand the web of life that surrounds and sustains us. Except for the white granular fertilizer we used in the potato patch, the gardening practices of my parents and grandparents had always been mainly organic. It was clear that the majority of the world was hooked on chemicals and had little knowledge, willingness, or patience for organic growing methods. Yet the deeper I went in my studies of nutrition, healing, and agriculture, the more it became obvious that what we were doing would not sustain us. In fact, it was—and is—killing us slowly.

Today, *sustainability* is a buzzword. Most people limit their ideas of sustainable living to the loss of rain forests, global warming, or dwindling energy resources such as oil and coal. They worry about whether or not they'll be able to afford their heating or cooling bills and bemoan the rising cost of food due to transport costs. But what about *us*?

When our food system goes down, we go down. We begin to suffer from degenerative diseases on a massive scale. We turn to drugs and surgery for help, but the cure is often worse than the disease. At present, we do not have a sustainable food system or sustainable medicine. Since these things are what currently sustain us to a huge degree, the question arises: are *we* sustainable?

What does the word *sustainable* really mean? In essence, it means the ability to continuously renew and regenerate ourselves. It means that we should not expect to become sick, weak, and helpless as we grow older. It means that we should expect to live to an age that is ten times our age at puberty, and longer if we want to work at it. It means that we should stop working for companies that destroy things and instead create work that sustains us and brings us joy. It means that we find our own personal balance point in life and honor that unique expression of self as part of our work in the world. It means that we are healthy in ourselves and at peace with our neighbors in the world; after all, there is nothing sustainable about war. I could go on and on.

The story of our unsustainable living is reflected in the story of a former neighbor's garden. He, too, grew up gardening the old-fashioned way, using organic methods, including sweat equity.

However, he wanted a big house on the water. To support it he needed a big income, so he went to school and then took a big job. He still loved gardening but began to be too busy to take care of it. When the "chemical" revolution began (euphemistically known as "the green revolution" back in the 1950s), he opted for artificial fertilizers and plant foods instead of compost. Later, bug and insect killers, which were really people-killers in disguise, were added to his garden routine.

Within a few years, his garden soil began to collapse, although he didn't know that at the time. All he knew was that his vegetables were overrun with destructive insects, so he used more insect sprays. Soon his garden was such a headache to care for that he stopped gardening altogether. It was easier to go to the grocery store or out to dinner. After all, he was a successful businessman. In his mind, he didn't need to garden any more.

"But something died when I let go of that garden," he mused, watching me plant. Now a bent old man living next to me in a small house on the shores of Anchor Bay, he spent his life taking pills and going to the doctor while dreaming of being able to convince his wife to agree to take down the tree in their yard so he could have enough sunlight for a small garden.

Although I didn't recognize it then, I was listening to the saga of our collective journey away from the soil and the sorry ending of that tale. When we let go of our relationship to the soil and hand the job of growing our food over to strangers, especially those on the other side of the world, we start down a road that leads to frustration and grief at many levels. When we look the other way and say, "I don't really know or care how you grow my food," we collaborate in an unfolding tragedy of massive proportions.

It is time we moved past the mind-set of "one size fits all," which, in agriculture, equates to "one farmer to feed all." When that farmer grows the same thing every year and chooses to sell it to food processors who turn it into food look-alikes, ruining the nutritional value in the process, we have no voice in the transaction. We are left to try to find something on the shelf that we can fill ourselves with, whether it has any nutrition in it or not.

Nutrition is king, when it comes to food. We would not expect to build and maintain a decent house without all the necessary materials and supplies, so how is it we expect to build and maintain a decent body/mind system without the necessary amino acids, minerals, vitamins, and enzymes in our food?

If we are not insisting on gardening and agricultural practices that produce high-density nutrition, then neither we nor the

growing methods are sustainable. Without top-quality nutrition, we end up physically degenerated, mentally fatigued, emotionally strung out, and empty spiritually. We lose our connection to the web of life and begin the slow amble of the walking dead that leads to the collapse we call death.

Gardening fosters a deep understanding of the interconnectedness of all life, and for that reason alone, it furthers our sustainability, strengthens our spirit, and transforms consciousness.

Sometimes the transformation starts early, as it did for me at the age of six. Sometimes it starts late, after retirement or when serious illness triggers a search for your personal truth and healing. Regardless, *the soul of the garden is its connection to the web of life*, and it is this connection that flows from the spirit of the garden to the spirit of the self, transforming all in the process, bringing wisdom, and teaching us about life.

My struggle to grow a few tomatoes on a hot deck taught me that there truly is a place for everything, and both people and plants do well when they are planted where they belong. We can survive or make do for a while in a foreign situation, but life unfolds with natural ease, grace, and bounty when we are in the right place, the right environment for us, with all of the people, things, and conditions the deep self cries out for.

I began to see that people are like seeds. Some are fat, some skinny. Some sprout quickly, some more slowly. Some produce big, lusty plants; some are wispy and frail. Some do well when planted together, while others must be kept apart. As with seeds, people's spirits are organic. They make space in their lives for their talents and skills to develop and grow naturally. They don't push the river of life as if hurrying is faster and better. These people have an innate sense of timing and recognize that Mother Nature is not the only one with seasons. The office has its seasons, families have their seasons, and certainly each of us has our own personal seasons.

The garden is a great place to learn self-discipline. Success will not be yours without discipline. The discipline of daily weeding and hoeing can be easy or difficult, whichever you choose. You can get out there "when you get around to it" on a hot July afternoon to suffer and sweat your way through the daily chores. Or you can get up at 6:00 am to do what must be done when the air is fresh and cool, insects are less active, and there is no feeling of anxiety that you are missing important phone calls. By 9:30 am, the sunshine is getting hot, and it's time to go inside, which is just about the time the business world is getting started in earnest. Voilà! You find there is a time for everything, both starting and stopping.

The old saying, "Make hay when the Sun shines," is a perfect example of learning to recognize opportunity. You can't till the garden in the rain, and it makes no sense to be repairing tools or shopping for trellising and gloves when the Sun is shining and the weather is perfect for weeding. You begin to see that "now is my chance" to do what must be done. Putting things off does not result in a payoff. And failure to recognize opportunity means you will never be able to follow your dreams, because you don't see when the door is open.

Watering a garden teaches you to nurture others. You begin to recognize when someone is wilting from pressure. You would never think of yelling at an eggplant when it needs something, so why would you yell at a person when they need something?

Harvesting is a critical test of whether or not you really want success. There is a perfect moment to pick corn; there is an ideal size for beans; there is a perfect time for picking tomatoes or peaches.

Collecting seeds for next year is akin to the assurance of being able to stay the course. Seed collection brings many gifts. There is the gift of observation. Which plant was early? Which had unusual and useful characteristics? Which was biggest? There is the gift of paying attention to small details. Collecting seeds teaches you to analyze what you have and recognize true possibilities for the future.

It is hard to say whether the gardener sustains the garden or the garden sustains the gardener. Deep within every true gardener there runs a river of spirit that sustains both self and the land. And yet the gifts that flow from the garden to the gardener—whether peas or peace—are miracles of sustenance and life. They are the gifts that appear in every sacred relationship. In the garden, the spirit of the gardener and that of the garden come together to nurture the gift of life and uncover the secrets that sustain, renew, and regenerate both.

About the Author

Penny Kelly is the owner of Lily Hill Farm and Learning Center in southwest Michigan. For thirty years she has been studying, researching, teaching, consulting, and writing about consciousness, perception, and intelligence. She is also a Naturopathic physician who gardens organically and teaches courses in organic gardening that are designed to increase sustainable living and food security in the City of Kalamazoo, MI. Penny is the author of five books and lead author of 14 eBooks on self-development written for the Ultimate Destiny Success System, and she runs two small publishing companies. She lives and writes in Lawton, MI.

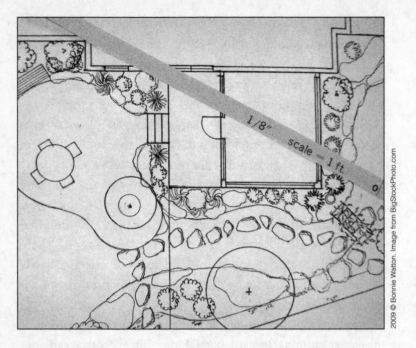

Building Your Wildlife Habitat

By Misty Kuceris

We see the beauty of the world around us, and we want this beauty to be part of our lives. So we build our homes, change the landscape, and all too often cultivate the land in the manner of Saturn, where rows are in order and structure exists. The flow of the Moon is forgotten, and our real neighbors—the animals and plants that attracted us to the land—are ignored or gone.

Enchantment exists in our lives when we are part of the world where animals and plants bring forth wood devas and fairies that bless our property with love and balance. By creating a wildlife habitat that supports nature on our property, we are once again in balance with the cycle of life.

The need to create natural habitats has existed for many years. Yet only recently has this need risen to our consciousness in such a manner that many resources now exist. Organizations such as the National Wildlife Federation (www.nwf.org) offer a program to certify home habitats. Wild Ones® (www.for-wild.org) provides guidelines for developing natural landscapes. And the Lady Bird Johnson Wildlife Center offers a Native Plant Information Network (NPIN) on their Web site (www.wildflower.org). Even local state programs can help people establish habitats.

What is in common with all these efforts—whether they are called building a natural habitat, eco-savvy gardening, or even xeriscaping—is that they call upon you to rekindle the harmony that can exist in this world without sacrificing planet Earth.

First, it's important to understand just what a wildlife habitat is. It's an area of land that supports local wildlife and improves the ability of natural plants to survive. In places where rivers, lakes, and bays exist, it improves the water quality through the decreased use of fertilizers and pesticides. In places that are arid and dry, a wildlife habitat ensures that the weather patterns and native flora don't change because too much moisture is added to the land from overwatering of lawns and attractions such as golf courses. In arid locations, such as deserts, it ensures that water is saved through the planting of drought-tolerate native plants. While wildlife habitats support local life and foliage, their real purpose is to keep the flow of the Earth in its most natural balance.

Creating a wildlife habitat in your yard doesn't mean that your property needs to look like the forest, although there are some people with so many trees on their property that they actually can create woodland gardens! Developing a wildlife habitat on your land also doesn't mean you have to take out all your lawn and replace it with what looks like unkempt weeds strangling the entryway to your home. You can create a beautiful wildlife habitat that works well in today's suburban setting and even educates others about the importance of a self-sustaining outdoor living space. All you need to do is study the various planting possibilities and hardscape enhancements that create a natural garden. With the right plants, you could even do this on your balcony or patio!

When you decide to begin, it's important to take a solar cycle (one year) to understand your land. You want to know where the Sun rises and sets throughout the year, creating light and shade on your property. You want to know how the light of the Moon brightens your yard at night. You need to know which are the east, west, north, and south sides of your home. Even the flow of wind currents is important. Perhaps an area that would be considered cold by other people's standards actually creates a warm, mini eco-climate in a particular location because of the currents that flow there. Or, you may have a windy perch of land that freezes even though it is on the south side of your home (considered the warmest side in the Northern Hemisphere).

Walk around your land, even during those times when the weather seems harsh. Perhaps you may find that ponds appear when the rain is hard. Or, perhaps you live in an arid location that spawns wildflowers during the evening when the land cools down, ones that are invisible to your eye during the daylight hours.

Once you understand your land, listen to its sounds, and feel its rhythm, you've actually tuned in to the cycle of life that exists around you. Now you're ready to start planning your wildlife habitat.

Developing a plan can seem daunting if you try to change everything on your property immediately. So, it's really best to start with baby steps. During the harshest time of the year, when you really can't go out and play, stay home and draw a map of your property, putting in your house, driveway, and any other permanent pathways. If you own your space and don't want to draw this map, you can use the plot of your land, which may have come with your closing documents when you purchased the property. Be sure to make the compass points on the land and even put in the wind patterns by using symbols such as arrows or wavy lines. (Think Sagittarius or Aquarius here.)

Divide your property into zones. Determining a zone is actually easy. One zone is the front of the house that has a lawn. Another zone is the front of the house that has a garden. If there are two gardens in the front yard, you now have three zones. The side yards are separate zones, as is the backyard. And, remember, any gardens in your side and back yards are also separate zones. Now just choose one of these zones for your first wildlife habitat.

If you live in a condo or apartment that has a balcony or roof, your zone is already chosen and all you really need to do is consider which plants can attract the birds or butterflies.

When you start, choose a zone that is small and doesn't require a lot of hard labor. Decide why you want the wildlife habitat. Do you want to attract birds or butterflies? Do you love the bats or flying squirrels that come out at night? Are you concerned about conserving water? Knowing your purpose will help you choose your plants and even your hardscape. And, what is hardscape? Hardscape can be a bench you place in the middle of a garden where you can sit and enjoy the view, a birdbath that provides water, or a rock that frogs and toads use to hide from the world and butterflies use to warm their bodies in the Sun's rays.

Deciding what to plant is not as difficult as it first seems. The National Wildlife Federation has divided the United States into several zones—Northeast, Southeast, Midwest, Rocky Mountains, Pacific Northwest, and Southwest—and has a list of their top ten native plants for each region. The Lady Bird Johnson Wildflower Center also divides the United States into several zones: Northeast, Mid-Atlantic, Southeast, Midwest, Rocky Mountain, Southwest, California, Northwest, and Canada. Both Web sites have maps of the United States so you can click on your state and get a list of native plants that are considered the best for your region. If you don't know what the plant looks like, some sites provide pictures of the plant as well as information on which type of wildlife that plant will benefit.

A beautiful and easy way to plan your garden is to get pictures of the plants and hardscape you want to place in the garden, a large piece of white paper or board, a pair of scissors, and tape or glue. Map out the garden and place the images in the spaces. Once you're done, put this board in a location where you can reflect and meditate on it until it's the right time to plant. As you meditate on this board, you may decide to make changes because the flow isn't quite right. Perhaps you didn't like the location of a plant or the hardscape. Or, perhaps you decided that you'd prefer a different plant in that location. If you find that you're making some changes, it's because you've tuned into the universal energy of the flow of the Earth.

During your meditations, use the cycles of the Moon to tune into the energy you've created while designing the board. As the Moon is waxing (increasing in light), meditate on what you have in front of you. As the Moon becomes full, pay attention to any dreams or images that come to mind. When the Moon is waning, you'll know if you have to make changes to the board or if your plan is the right one for you. When the season is right, it's time to go outside and put this habitat landscape into action!

There is a general rule for all perennials, trees, and shrubs. The first three years go to building a good root system; it may take a while before flowers and fruits appear. This is especially true with native plants, in particular trees and shrubs. In some cases, it may take up to five years before your trees and shrubs bear the flowers and fruit you want them to. So, don't be discouraged if your plant looks a little scrawny the first few years. Also, while native plants do better in an area and need less water and fertilizer than the nonnative species, this only occurs after they are well-established. As such, you need to make sure to water them well during the first few years, until their roots are firmly established.

To assure that tree and shrub roots take, plant with the Moon in a water or earth sign and, if possible, forming a good aspect to

Jupiter. To assure that the roots of perennials take hold, plant with the Moon in a water or earth sign and, if possible, forming a good aspect to Mercury. Whenever you plant, try to have the Moon increasing in light (waxing).

As you watch your garden grow, do more than just keep an eye on the weeds. Take time to remove them because weeds can increase the risk of insect damage to your plants. While weeding is always best done with the Moon decreasing in light and in its fourth quarter, you really won't be able to wait a week or two between weedings. You'll need to be vigilant and do it whenever necessary. So, pay attention to the Moon and try to weed when it is in a barren sign, such as Aries, Gemini, Leo, Virgo, Sagittarius, and Aquarius. Keep an eye on your potted plants as well, because weeds will grow in them just as they do in the ground soil.

As the seasons pass, your native garden will grow and fill out. As it fills out, you'll experience less work and greater pleasure in seeing what living creatures come to your paradise.

Some Native Plants to Consider

While various organizations divide the United States into regions, it's important to realize that even different parts of a particular region may have more than one type of climate. This is especially true if you live in the Western portion of the United States. For example, if you go to the Web site of the Native Plant Society of New Mexico (http://npsnm.unm.edu), you'll read that they have five different regions just in that state, ranging from desert to mountainous. Also, while native plants in one region will grow in another region because the zone is the same, that plant isn't always considered native in both zones. It is always best to contact your local university or native plant society for a detailed list of plants native to your region. For an online list of regional native plants, see PlantNative (http://www.plantnative .org/).

A colored map of the numbered zones can be found at the National Wildlife Federation's Web site at http://www.nwf.org/gardenersguide/gardenzone.cfm.

Northeast

The Northeast states tend to be found in zones 3 through 7 and located in the northeast portion of the United States as far south as Virginia and as far north as Maine.

If you are interested in planting a tree, you might consider the river birch (*Betula nigra*), since the seeds attract birds, or the persimmon (*Diospyros virginiana*), since this is considered a butterfly larval host. The serviceberry (*Amelanchier*) is a wonderful shrub that provides edible fruit and the winterberry (*Ilex verticillata*), another shrub, also attracts birds and mammals and provides good winter shelter. Just remember that the winterberry has both a male and female shrub so you'll need one of each in order to produce berries.

When you want to attract hummingbirds to your area, consider the cardinal flower (*Lobelia cardinalis*), a gorgeous red flower that actually needs hummingbirds for its pollination process.

Southeast

Southeast states are found on the Atlantic coast of the United States and go from North Carolina to Florida and as far west as Texas. They are found in zones 7 through 10.

Trees to consider, especially for butterflies, are the Mexican redbud (*Cercis canadensis var. mexicana*) and the black cherry (*Prunus serotina*). The shrub woolly bee bush (*Aloysia macrostachya*) is also another good butterfly larval host while the serviceberry (*Amelanchier*) provides berries and cover for birds and mammals.

Wonderful plants for zone 9 include the wild blue sage (*Salvia azurea*), which is a butterfly larval host. In zone 6 or 7, consider bee balm (*Monarda didyma*), which attracts bees and hummingbirds.

Midwest

The Midwest region stretches from Ohio to North and South Dakota and can go as far south as Nebraska or Kansas. The primary zones for this region are zones 2 through 6. The river birch (*Betula nigra*) and black willow (*Salix nigra*) are both great hosts for butterflies. And, as long as you are willing to plant a male and female shrub, don't forget the winterberry (*Ilex verticillata*) to provide shelter and berries for birds and mammals.

Southwest

The Southwest is actually a tricky region when considering native plants because its zones go from 5 to 10 and can change even within small areas of the state. This can be New Mexico and Arizona. However, the region could also include parts of southern Colorado, southern Utah, and southern California. If you live in this area, it really helps for you to contact your local native plant society for specific information.

The skunkbush sumar (*Rhus trilobata*), a shrub, and the blue paloverde (*Parkinsonia florida*), a tree, provide food for both birds and mammals.

Rocky Mountain

The Rocky Mountain region is another tricky area when you are looking for native plants. Again, it's best to check with your local native plant society for specific information. The states found along the Rocky Mountains go from zone 4 to zone 9.

The narrowleaf cottonwood (*Populus angustifolia*), a tree, provides shelter for birds who can build nests in the soft wood. The plant bee balm (*Monarda didyma*) attracts bees and butterflies, and skyrocket gillia (*Cilia aggregata*) attracts hummingbirds.

Pacific Northwest

The Pacific Northwest region is primary zones 7 through 9, although some zone 6 plants will live in regions close to the mountains.

The Douglas fir (*Pseudotsuga menziesii*) is best known as a Christmas tree, yet it provides shelter and food to birds and mammals. The red osier dogwood (*Cornus stolonifera*) is a beautiful shrub that loses its leaves in the winter, leaving gorgeous red branches in your landscape that will provide berries for birds. For plants, consider one of the Penstemon group of flowers or scarlet gillia (*Ipomopsis aggregata*), which attracts hummingbirds.

About the Author

Misty Kuceris has worked as a plant specialist over the last several years for various nurseries in the Greater DC metropolitan area. As a plant consultant, she meets with homeowners to assess their property and gardens. She also lectures at garden clubs and senior citizen centers giving guidance and advice on how to create healthy home gardens and lawns. At the time of this printing, Misty's gardening Web site is under construction. However, you can reach her at Misty@ EnhanceOneself.com with any questions.

Urban Gardening: How to Garden in Small Spaces

by Pam Ciampi

What I've been hearing a lot these days about gardening sounds something like this: "I'd really love to have a garden, but I just don't have the space, the time, or the money to do it." Between the rising population, the cost of living, and all the new electronic inventions and devices we have to monitor today, our lives seem to be simultaneously shrinking and escalating out of control. We have less space to live in, less time to enjoy the space, and less money to enjoy it with. If we take a tip from Gandhi, "To forget how to dig the earth and tend the soil is to forget ourselves," nothing short of a green revolution will be able to help us find our balance in these crazy times. It doesn't matter if you are one of the many who don't have the place, time, or the money to plant an old-school type of garden—all you need to join the green revolution is a desire to get back in touch by growing beautiful plants and healthy inexpensive food in whatever space you have. Small is beautiful and green is the new gold, so start to plan your urban garden today!

We are living in new times that call for new solutions. Urban gardening is a new mindset as well as a different method of growing plants: it is a viable solution for people who like to think outside of the box. It gives people who are looking for ways to garden in a cityscape an accessible, affordable solution that is creative but not labor intensive. As a method, urban gardening has no rules. It uses whatever soil and containers are available to grow plants and vegetables. Usually this means planting a container garden either inside or outside on a patio, a balcony, in window boxes, or on a rooftop. Urban gardening can also mean landscaping with edibles or starting a community garden in an abandoned lot.

Here is a list of some things to consider to get you started on making your urban garden grow: space, light, containers, plants, and last, but not least, how to use the Moon to keep you in touch with natural cycles and rhythms of Earth. On a practical level, finding out where the light is coming from will tell you the right space to set up the garden. Your choice of containers will determine what kinds of plants to grow. On the sacred level, being aware of the positions of the Moon can tell you not only what type of gardener you are but the best times of the month to plant.

Gardening by the Moon

There's a joke about having a perfect marriage that also tells how you can use the zodiac to help your garden grow. It goes like this: "my wife's a water sign; I'm an earth sign. We get along so great because together we make mud." It's funny but it's true—wet earth makes the best medium for any seed to grow, whether it's for a plant or a relationship. The zodiac sign your natal Moon is in tells about your ability to "make mud," which determines your natural gardening type. People with green thumbs have their natal Moons in signs that are fertile and have high nurturing abilities, such as Cancer, Pisces, Taurus, and Virgo. People who are born with a black thumb have their natal Moons in signs that are not naturally warm and fuzzy, such as Capricorn or Scorpio. These types of gardeners do very well with no-fail types of plants. No-fails are plants that will flourish without a lot of care—carrots, radishes, and peas, as well as succulents like aloe vera or cactus.

You can also use the Moon's phases to help your urban garden grow. Here's a simple way to determine the phases: if you can look outside and see the Moon, hold up your cupped hand to its shape. If you can fit the curve of the Moon in your right hand, the Moon is waxing. If it fits into your left hand, it is waning. If city lights or buildings block your view, go online to a Moon site or consult an almanac, such as this one, or an astrological calendar to determine which phase the Moon is in. Plant your seeds during the waxing phase when the Moon is growing (the two weeks from New Moon to Full Moon), and plant root vegetables during the waning phase when the Moon is getting smaller (the two weeks from the Full Moon to the New Moon). By using these two easy rules you can give your garden an added boost by getting on track with natural cycles.

Space and Light

The next thing is to decide where you want to set up your urban garden. Keep in mind that small is beautiful and that, with a little imagination, any area can be transformed into the perfect garden. A concrete patio? Terrific! A tiny porch? Even better! A balcony, window box, or even an abandoned lot filled with weeds? A nice challenge! I recently brought home a funky, rusted-iron window box from Mexico, nailed it under an outside west-facing window, lined it with moss, and called it an herb garden. It was a quick and inexpensive way to keep my kitchen stocked with fresh basil, chives, and parsley all summer. If you don't have any outside space that is available and you are limited to the inside, feel free to use a table, countertop, windowsill, or anything else you can come up with. Urban gardening is all about thinking outside the box. The most important thing to remember is that plants need a good source of sunlight.

Now that you have some possible spaces in mind, it's time to find out what kind of light you have available. The rule of thumb is that if you want to grow vegetables, look for the sunniest spot. If your have chosen a space that is dark and doesn't have access to a fair amount of sunlight, you will be limited to growing mushrooms, sprouts, or houseplants such as ferns. Most plants need lots of direct sunlight in order to grow.

To figure out what exposure your urban garden space is getting, start by observing where the morning light falls. Morning light comes from the east and provides good light for most plants; the only drawback is it only comes in the morning and doesn't last all day. If you are blessed with light for most of the day, you get the green thumb prize because you have a southern exposure—the best light for growing. If there is only light in the afternoon (western light), it is excellent, as it is hotter and lasts longer than morning light. If no direct sunlight falls on the space you are considering, your space is north facing. Spaces that face north have terrific light if you are an artist, but they should be avoided for a container garden.

Containers

The next step is where you can really get crazy: containers. There is only one rule here and that is, if it can hold dirt, it's a container! The most important thing is to be sure you can poke a few holes in it so the water can drain and the dirt doesn't stay soggy. Here's the time to think "recycled" and let your imagination be your guide. Besides the obvious clay pots, plastic jars, wooden barrels and boxes, anything is up for grabs. I've seen containers made out of old boots, tires, buckets, bathtubs, wheelbarrows, watering cans, unused toilets, and plastic bottles. If you are really strapped, you can always buy a bag of dirt, lay it on its back, poke a few holes in the plastic and you are good to go.

Plants

Now that you've determined your urban garden's location and have your containers ready, it's time to decide which plants are best for your urban garden. Keep in mind that you should always choose the right plant for the right container. A tomato is not suitable for an old shoe but a parsley plant might fit perfectly. Please do not feel limited by these suggestions; add your own! And do not worry even if you don't have access to any dirt. After listing some edible plants that like dirt, I've also included sprouts, which are soil-free.

There are many different types of plants to grow in your urban garden, including herbs, roots, vegetables, and dwarf fruit trees. The easiest herbs to grow are parsley, sage, rosemary, and thyme. I get a wicked pleasure out of potting mint, one of the most invasive plants in the herb family. Basil and dill, chives, and curry also seem to thrive in containers. Carrots, garlic, ginger, and radishes are some no-fail roots that are guaranteed to grow. Make sure they have good drainage and put them on an indoor windowsill with a sunny east or west exposure. For maximum sunlight, the smaller varieties of vegetables—like cherry tomatoes and bush beans as well as peppers—can flourish in outside containers. As for fruit trees, the dwarf varieties are the only kind that will thrive in larger containers such as half barrels. I once grew a dwarf mandarin orange that produced fruit indoors for years by a south-facing window in Vermont.

Last on the list of plants for the urban gardener is sprouts, the answer to the indoor gardener's prayer. Sprouts, otherwise known as germinated seeds, are the wonder food of the urban garden. An ancient micro-food, sprouts pack more nutrition for their size than any other vegetable because they contain so many essential nutrients and enzymes. They aren't picky and don't need any sunlight (note to those with northern exposure). The best place to grow sprouts is in a glass jar near your kitchen sink. All you have to do is put the seeds in a jar with an inch of water for a day and then rinse them once every day. Before you know it you have fresh food at your fingertips! Start with mung beans (yummy in Asian dishes) or alfalfa seeds that go great on a salad. Sprouts are tasty, easy to digest, and always available even in the dead of winter.

About the Author

Pam Ciampi is a full-time professional teaching and consulting astrologer. She served as president of the San Diego Astrological Society from 1998 to 2005 and is currently president of the San Diego Chapter of the National Council for Geocosmic Research (NCGR). She holds a Certified Professional Astrologer diploma from International Society for Astrological Research (ISAR), has been a faculty member of the American Federation of Astrologers (AFA), and is a longtime member of the NCGR. She has completed the ISAR Ethics Training Course and has also served on the Ethics Committee. Pam is an amateur but enthusiastic gardener. She has published three annual almanacs entitled Gardening by the Light of the Moon, *based on astrological gardening practices.*

Biophilia Hypothesis

By Carole Schwalm

The biophilia hypothesis (introduced by E. O. Wilson) means the love of life and the human need to relate to other life forms. In other words, we need wildlife as much as wildlife needs us. And they need habitats. Humans destroy wildlife habitats and lessen the blessings of biophilia interaction. But humans can also create and re-create habitats.

With habitats, we aren't talking short, designer-type grass. No! Clipped hedges? Not! A dandelion-free or completely weed-free lawn? I don't think so. Large patio or five-car concrete driveway? Nada, nix, nil! Pristine conditions free of brush piles or dead branches or twigs? No, thank you very much. Everything done military style in straight lines? Think again.

Breathe, oh great and wonderful human being, and loosen control of your yard. Picky, or do you have picky neighbors, friends, or family? Your answer to comments on your natural lawn is that you are building an ecosystem or a naturescape in your yard or apartment balcony. If blank stares continue, say it's a re-creation of an old-fashioned cottage garden.

Through the eighteenth century, a yard was natural. Whatever grew, grew in a balanced land of thriving flora and fauna. Voilà, along came the manicured look at the Palace of Versailles—The Lawn. The European rich, the elite, wanted to align themselves to royalty and separate from the common folk, so they emulated palace formality. British colonists brought the lawn to America. It took one hundred years, or into the nineteenth century, for the middle class to adopt strict expanses of green.

The formal lawn was mowed. Ergo, the invention of the lawn-mower through history, from hiring people to use scythes, to push mowers, to today's lawnmowers that pollute as much in an hour-long mowing session as a car does traveling 350 miles. In the United States, more than fifty-eight million gallons of gasoline a year are put into lawnmowers.

All those clipped lawns lessen habitat. Wildlife likes longer, more natural grass, not the nonnative designer grasses. Habitats flourish with native grasses that are allowed to grow longer, needing less mowing. In the spring, unmowed grass (between 4 to 5 inches) hides wildlife babies in what fauna trusts are protected nests.

An eco-lawn of native grass comes in seed mixes that attract wildlife, especially butterflies. Since it isn't nonnative designer grass of little value, it takes less fertilizer to grow. It takes less water. In the United States, 30 to 60 percent of fresh water is used just to water our lawns.

Groundcovers are alternatives to vast expanses of grass. They don't need mowing, need less water, and provide fruit in return. Many like shade. Before the 1950s, clover was a lawn seed. In comes marketing, and a manufacturer invents a chemical that kills clover, but leaves the grass. Marketing declared clover an ugly weed, even though it was green. Clover made a lawn that required less water and mowing, but wildlife likes it. Bees and bunnies love clover! Unbiased birds even love crabgrass.

Clipped hedges? Get too fancy with your clipping and wildlife loses a place to hide. This is especially important close to your bird feeder and water supply. There are fewer branches to perch on, not to mention fewer nesting spots. You may want to clip blatantly dead branches, but leave dead branches close to the inner core, especially through the fall and winter months. Insect life still lives in them and birds eat insects. Shrubbery close to the house or near fences and trees provide warmth to winter wildlife.

Shrubs and hedges provide both shelter and food. They fill out corners of the yard, and they don't mind being crowded in groups of varying plants. Shrubs become even more valuable if they provide berries and seed-filled cones.

Weeds? Some cities and towns, in light of the above lawn scenario, adopted weed laws with the idea that weeds lowered property values, brought rats and mice, bred mosquitoes and pests, and presented a fire hazard. And the answers: They don't lower values if they are accepted and welcomed through habitat preservation (you can also put up some type of border to avoid offending other people with your ecosystem); rats and mice prefer manmade buildings and garbage; mosquitoes like standing water, and weeds bring birds who eat insects and pests; native plants, even weeds, absorb water easily, remain moist, and are less of a fire hazard. How about the allergy argument? Weeds and native plants are insect pollinators, not air pollinators. If you are allergic to a certain weed or native plant, avoid that species, but you can certainly have others.

Unfortunately, the common way to attack weeds is through the use of insecticides. In other words, a big no-no when it comes to creating a natural wildlife habitat. Birds eat insects that ingest insecticides. A poisoned insect poisons a bird. Killing insects reduces the food supply for birds as well. The Audubon Society estimates 67 million birds are killed by pesticides worldwide each year.

The nonuse of pesticides is better for you, your children, and your pets—think about the chemicals they are exposed to as they roll around on your weed-free lawn. Insecticides and weedkillers have been found to contain things like Agent Orange, nerve gas, and artificial hormones. Toxic vapors can remain active for years. Insects are best controlled by other insects. That ugly old caterpillar chewing a few leaves is a butterfly in progress.

Wildflowers multiply on their own. If the climate isn't right, they won't flourish. If there is a drought, they won't come up this year, but perennial species will be back the following year. You'll have more of what exists happily in your yard. Remove those plants that don't supply food and replace them with ones that do. There is no law against mixing wildflowers, even a few things that some consider weeds, such as dandelions and milkweed.

Patio or concrete? Patios and concrete are dead space. Nothing grows. It doesn't mean that you shouldn't have it, just keep these areas small. If it's already there, think about adding containers or flower boxes around the edges. Stone walls or dividers can be wildlife homes or shelters.

No brush piles or dead branches? You don't have to let the brush piles grow to massive sizes, but smaller piles create hiding places for worms and bugs, and the birds will find them. Birds use twigs and even small weed stems for their spring nests. A little hummingbird makes a nest with moss down, dandelion fluff, and spent flower blossoms sometimes bound together with spider webs. Leave them some of the above nesting materials, and the hummingbirds will pollinate your flowers and eat insects in return.

Dead branches can also include seed heads. If you deadhead the flowers through the spring and summer, you encourage blossoms. Leave the seed heads on at the end of the season and let birds harvest the seeds for you. It goes against feng shui ideals to have dead trees or branches around, but a few in a minor location won't hurt. There are insects in the trees for birds to eat. Squirrels do what squirrels do in them. If it bothers you to have branches and piles of leaves, put them around the edges of your yard and plant in front of them. Plant vines around a dead tree.

Regarding those straight lines: A visit to a wild meadow is the best example of wildflower growth. Every little flower isn't lined up in reds, yellows, and whites. There are sprinklings of colors, and no straight lines. The same principle applies in the yard. Plant in layers or mimic a natural habitat. That means low-lying, medium, and tall. The latter two serve as lookout areas for wildlife.

Have transition areas. Several of the transition areas should include water, with shrubs or trees around it for safety. Make sure cats can't hide in them! Birds stay around water in the wild, but they are wary of anything more than two or three inches deep. I use large flowerpot saucers with slightly taller rocks in the middle for main drinking sources and long flower box saucers for bird-bath purposes. It doesn't have to be fancy, birds aren't snobs. They just need water to drink, wash, and use to make mud for nests. It's kudos and kisses if you get the water to move (waterfalls or ripples) and more of the same if you remember to deice it during the winter.

To become an ecosystem, your habitat can't just attract birds and butterflies. It needs to attract other things, and, in light of loosening the attitude of lawn control, get ready to exist peacefully with toads, frogs, lizards, and members of the kingdom that includes birds, mammals, reptiles, amphibians, and insects as well. You'll get a beetle. Along comes something that thinks a beetle is a fine dining experience, a snake. Hey, that's natural rodent control.

Your areas should be a source of food 24/7, 365. Look at seasonal growth, because something provides food all the time.

Native sumac, for example, holds fruit long enough to provide sustenance for migrating birds.

Habitat-friendly Flora Friends

Pollinator attractors (and we need those little pollinators for the "birds and the bees" process): Begonias, hollyhocks, hibiscus, lantana, morning glories, milkweed, yarrow, stonecrop, gay feather, ocotillo, salvia, desert marigold, phlox, Queen Anne's lace, columbines cacti, thistle and thistle seed, mesquite, rabbit brush, acadia, honeysuckle, sumac, creosote, lilac, holly.

Berries and fruit: Juneberry, elderberry, huckleberries, wild grapes, cherries, blueberries, crabapple, blueberries, blackberries, raspberries, yews, plums, hawthorne, roses for rose hips, gooseberries, snowberries. (Make sure the plants aren't classifed as "sterile." Sterile plants were created for people who don't like flower droppings on their lawn.)

Insect attractors (which also attract insect eaters, i.e., birds): hackberry, slippery elm trees, dogwood. Fruit trees in blossom attract insects.

Seeds for the seed lovers: ash trees, asters, birch trees, coneflowers, goldenrod, legumes, maple trees, native grasses, nut trees, oak trees, sunflowers. Conifers are sources of seed. Also sassafras, bee balm, lavender, salvia, rosemary, dill and fennel, bluestem and switchgrass, clover, native grasses, sagebrush.

Ground covers: wintergreen, wild ginger, moss, ferns.

Contact the references below for information on the sources of plants native to your area. You can order seeds and even participate in a seed exchange, if you like. A good local nursery is another good resource for information on native species. Make sure that their plants are "nursery propagated" from seeds. Consider talking to builders just starting to level land. Ask if you can collect seeds or native plants that are ready to die in the name of progress. Be sure to get permission; do not trespass on the property, even if it is an open field.

It is more cost effective to make changes slowly. You don't have to get rid of all your exotic plants at once. Just add native plants to the mix. Garden resources will tell you which types of plants to welcome and which to avoid. And you don't have to say goodbye to your asters, petunias, or marigolds—insects, hummingbirds, and butterflies like them!

About the Author

Carole Schwalm lives in Sante Fe, New Mexico. She has contributed self-help articles and horoscopes for many people through America Online and other Web sites. She currently provides a variety of astrological work for Astrocenter.com.

Online Resources

The Fish and Game Department in any state in the United States lists the top ten native plants on their Web site. Simply use a search engine to find your state's site. The National Wildlife Federation (www.nwf.org) has similar lists in their Certified Wildlife Habitat–Food section (www.nwf.org/backyard).

Enature.com has maps and lists of varying wildlife and endangered species on its Web site.

The National Audubon Society's Web site (www.audubon.org) has information on birds and habitats.

The Web site of the organization Wild Ones® (www.for-wild.org) provides information on native plants and natural landscapes.

Books

Anderson, Katharine with Tom Carpenter, Justin Hancock, and Ann Price. *Wildlife Gardens*. Minnetonka, MN: National Home Gardening Club, 1998.

Bormann, F. Herbert, Diana Balmori, and Gordon T. Geballe. *Redesigning the American Lawn: A Search for Environmental Harmony*. New Haven, CT: Yale UP, 1993.

Jenkins, Virginia Scott. *The Lawn: A History of an American Obsession*. New York: Smithsonian, 1994.

Gardening by the Phases of the Moon

By Maggie Anderson

In mid-February 2008, I drove through the northeast Iowa countryside and was startled to see dozens of farmers in their fields on tractors, plowing through several feet of snow! I'd lived in the area for most of my long life and had never seen such a thing. *They surely can't be planting corn now!* I thought. *Do they have really bad cases of cabin fever?*

Later that day, I was relieved to hear on the local ag-news radio report that my rural neighbors were digging trenches. When the spring thaws came, these trenches would collect water melting off from the heavy snows (seventy-six inches) of the winter of 2007 to 2008. The farmers' actions were rational, not mass hysteria brought on by an extremely long winter (there was plenty of that among the townies like me).

As an Iowa native, I knew farmers don't plow and plant in February, and I knew it was February because my calendar told me so. It's a solar calendar that tracks the path of the Sun through the zodiac each year. Since the Sun is the source of heat, farmers and gardeners plan their planting days by the solar calendar, with keen attention to planting "zones"—the farther south in latitude, the earlier the seeds can go into the ground.

Solar and Lunar Calendars

Not all nations base their calendars on the movements of the Sun. Throughout history, the Moon has been associated with the movement of ocean tides and rainfall, so agricultural traditions grew up around tracking the 28½-day cycles of the Moon. The first calendars were lunar, including those of the Hebrews and Chinese. Most of these originated in locations close to the Equator, where temperatures were constant but rainfall was critical.

Some countries have continued the lunar calendar tradition, including Islamic nations. Others, with a bow to the modern Western industrial world, publish calendars that note movements of both the Sun and Moon. These recent adaptations are known as "lunisolar" calendars. In the West, it is primarily almanacs that track the Sun and Moon as regulators of both temperature and precipitation.

Why Gardening by the Phases of the Moon Works

Because calendars in the United States are based on the cycles of the Sun, we are less familiar with the cycles of the Moon. Thus, gardening by the Moon's phases or arranging other activities to coincide with her cycles seems unnecessary. Our agricultural ancestors, however, having tracked the ocean tides and rainfall for thousands of years, embraced Moon phase gardening. By the time

Pliny the Elder (AD 23–79), an ancient naturalist, collected all known natural science in his work, *Naturalis Historia*, knowledge of planting, growing, and harvesting by the phases of the Moon was an integral part of popular culture.

The overriding theory of the ancients was that it is best to coordinate all earthly affairs with the cycles of nature. Even close to the Equator, certain months of the year will not produce good crops because of excessive heat. Particular phases of the Moon became associated with rainfall, which is critical to a good growing season. The New and Full Moons were associated with "earth tides." This theory proposes that Luna not only moves millions of tons of ocean water with the tides, she also pulls moisture out of the air and into the ground at the New Moon. Full Moons draw water up from the ground closer to the surface, where it can provide moisture to plants. These theories have been confirmed by modern science. French Polynesia now hosts the World Data Center for Earth Tides, which collects tidal data from around the globe.[1]

It's Easy: Just Remember "D-O-C"

The first challenge for gardeners who want to realize the benefits of planting by the Moon is to be able to identify her four phases. The easiest way is to step outside after dark and take a look. You'll see one of the images described below:

• The first quarter or New Moon period of the waxing Moon is a slim crescent that resembles the right side of the capital "D."

• The second quarter of the waxing Moon looks like someone has taken a yellow crayon and colored in the "D." As it increases from first quarter to Full, the backside of the Moon becomes filled in too, until it becomes . . .

• The third quarter or Full Moon period of a waning Moon, the round "man in the Moon," a perfect "O."

• Next is the fourth quarter of the waning Moon. As the Full

Moon decreases, the right side diminishes, and Luna looks like a capital "C." The Moon becomes slimmer and slimmer until it disappears and then turns into a "D" again.

If it's too cold to go outside for a nighttime peek at the Moon, local newspapers report current Moon phases in their weather sections. There are also online sites with this information, such as this one maintained by the U.S. Navy at http://tycho.usno. navy.mil/cgi-bin/vphase-post.sh and a simpler version at http://aa.usno.navy.mil/data/docs/MoonPhase.php/

How to Plant in Harmony with the Moon

Assuming that you'll also pay attention to your solar calendar and planting zone, identify the months(s) in which you want to plant your garden. Then arrange your seed packets into three piles:

Annual plants that produce seeds on the outside of the edible veggie, like brassicas and greens. Plant these seeds or seedlings during the first quarter or New Moon, along with annual flowers. You can sow the seeds in flats during one first quarter Moon and put them under a grow light indoors or an outdoor greenhouse.

Wait approximately 28½ days, and the Moon will be new again and the ground ready to receive your seedlings.

Annual plants that produce seeds on the inside of the edible veggie, such as tomatoes, peppers, eggplant, cucumbers, squash, and pumpkins. Plant these during the second quarter Moon. The seed-to-seedling routine is similar to that for first quarter plants, with the same timing.

Annual plants that produce the edible part of the plant underground, including potatoes, beets, and carrots. In addition, all perennial flowers and vegetables that must establish a strong root system can be planted in the third quarter when the Moon is waning. Seeds for these are usually sown directly into the ground, unless you are following a bio-intensive gardening plan; then start seeds in flats and transplant outside about 28½ days later.

Also plant shrubs and trees by the Moon phase. Plant in the third quarter Moon, from Full Moon to the fourth quarter Moon. Weed and mulch. Then weed some more.

Refining the System

When you first begin planting by the phases of the Moon, you might want to simply pay attention to what quarter Luna is in. Once you become familiar with that, the system can be refined and made even more successful by planting under a hospitable sign of the zodiac. Two earth signs and all water signs have traditionally been more compatible to plant growth than the others.

Earth Signs: Taurus and Capricorn. Herbs, flowers, and vegetables love to be planted under a New Moon in Taurus, and potatoes adore going into the ground when the Moon is full and in Capricorn. Remember Virgo is a time for harvesting rather than planting.

Water Signs: Cancer, Scorpio, and Pisces are the best of the best for planting. However, since the Moon in water signs produces rain, you may need to run outside during a lull in a storm to plant.

Having your beds prepared before the planting time will make it easier to incorporate water-sign planting into your routine.

Fire Sign: Only Sagittarius is good for planting onions and garlic—they'll be nice and hot if you do. Forget about gardening in Aries and Leo, except when you need inspiration—read your garden catalogues when the Moon is in one of these two signs.

Air Signs: These are generally too dry or windy for gardening. Your seeds will scatter and the whole time you're outside, you'll wish you were inside reading a good garden catalog.

Llewellyn's Moon Sign Book has laid out all the information you need to garden by the phases of the Moon. As you keep records of your success year after year, you'll become more aware of how reliable this method is. Your plants will be stronger, attract fewer insects, survive inclement weather, and produce larger harvests.

The system is not as complicated as it might seem at first. The key is to prepare your garden beds thoroughly and as early as possible, at a time of decent weather. Mark off your rows for quick planting and wait for the right phase of the Moon and the correct sign of the zodiac. Then when the stars and planets align, it will take only a few minutes to pop those seeds into the ground and cover them up.

Happy Planting!

1. The International Center for Earth Tides. The Royal Observatory of Belgium. Period 2003–2007. By B. Ducarne, ICET Director.

About the Author
Maggie Anderson is an astrologer and gardener who lives in Marion, Iowa. Her 2008 gardening adventure was to plant a "salad bar" outside her back door in a raised bed made from a horse trough.

Dry Gardening

By Janice Sharkey

Dry gardening is nothing new. If I had to choose to reduce one of the three crucial growing elements—light, water, or nutrients—I would opt to cut water every time. Water is essential, but *how much* is essential is the question. Although up to 90 percent of leaves and stems in plants are made up of water, how much water a plant necessarily needs varies. To see plants that thrive in hot, arid conditions, you need to visit South America, southern United States, South Africa, Australia, and the Mediterranean. The flora and fauna that survive in these climatic zones do so due to cleverly adapting to maximize every drop of water.

Now we are experiencing an increasingly volatile world climate. Increasingly hot summers in temperate climates have led to semi-drought conditions, with gardeners being banned from using hoses. We need to learn lessons from those plants that survive the arid soil and adapt our garden in an increasingly dynamic climate.

Drought-loving Plants

If you took a handful of soil from a typical garden and tested it for nutrients, you would find it was poor and very free-draining. Long exposure to baking sunlight causes water to evaporate within the leaves of plants, and the structure of the soil is likely to be poor and dusty. Few plants can survive these harsh conditions and so they need to biologically adapt to have any chance of growing.

Most drought-loving plants actually thrive in poor soil. If they have growing conditions that are too easy, they tend to become soft, which can make them susceptible to disease. Plants that like hot, dry conditions develop harder stems that are more able to withstand cold snaps when temperatures drop.

So how can you identify a dry-loving plant? Be aware that most drought lovers are gray and silver-leafed plants, such as artemisia, lamb's ears, brachglottis, lavender, and cotton lavender, to mention but a few. They thrive in dryness due to their silvery color, which reflects heat away from the plant. Peer into the plants and you might see very fine hairs. These hairs trap moist air close to the leaf's surface, creating a microclimate to protect the plant from scorching or shriveling. The trapped moisture also means the plant needs less water via its roots from the soil. Under the soil, where the roots are, other beneficial things are happening—the hairs on the roots are multiplying, allowing the plant to take up more moisture from the soil.

Most drought-loving flora have leaves that cover a small surface area (take gorse, for example, with its tiny needle-like leaves) or small leaves, as on lavender or rue. Just look at dianthus pinks with their gray narrow leaves or the gray jagged ones on a globe thistle that are two fine examples of nature creating a defense mechanism against the rigors of baking summer. They cover the maximum amount of space without covering a large surface area, avoiding the loss of vital moisture.

Sap within a plant plays its part in conserving moisture and helping plants stay alive. Often aromatic plants are excellent at this because they have a highly concentrated sap that is drawn from the soil and stored until needed—a bit like a camel stores fluids in its hump.

Herbs like sage, rosemary, and lavender also protect themselves from excessive heat and drought with their fine needle-like leaves—each leaf hangs, creating shade for the next leaf, so only a few are actually exposed to the rays of the Sun (this is similar to cacti, which cast shadows over their growth to shade against the sunlight). Cacti and other succulents store water in their thick fleshy leaves. Some plants have tough, leathery leaves that act as protection (as in the genus *Bergenia*) or a waxy coating (as in the genus *Euphorbia*). And the hairs on geranium leaves help it to be semi-drought-resistant.

One boon is that drought-lovers have wonderful flowers that are bright and act as a magnet to pollinating insects. Some will only open their blooms when the Sun is shining since insects tend not to appear in the rain. These plants flower and go to seed quickly but often repeat flowering frequently. Livingstone daisies, with their semi-succulent stems and leaves, and members of the genus *Osteospermum* are two such flowers that open and close with the Sun. These plants love to act quickly in response to the shortage of moisture. By deadheading them and sending the plant a message to repeat the flowering, you'll prolong its blooms.

Protect Plants from Dry Spells

As gardeners, we can try to maintain the water in the soil as nature intended, by digging in lots of organic materials that hold moisture and add wonderful nutrients to give a boost to the plant. Giving a plant healthy soil brings many benefits, including providing nutrients to help it fight harsh conditions. Earthworms are a good indicator of soil condition—worms like a balance between moist but not too boggy conditions. They will turn over, aerate, and fertilize the soil. Their presence higher up the soil strata helps the soil maintain a balanced temperature that gives an insulating effect. After you've given your plants organic materials in the soil, it's time to bed the plants down.

Mulch Mulch Mulch

By "bedding it down," I don't mean the plants are going to sleep, rather that they are being protected by a blanket that retains moisture. Mulching saves time for the gardener, and plants love growing through it. The best time to lay mulch is in early spring; you should have prepared the soil and watered the plants well before applying your mulch.

There are various mulches to suit your growing space and budget. The best mulch has a coarse texture that allows air and water to penetrate but also excludes light to stop weeds from germinating. A finger's length of depth is enough to stop light from getting through, but this much mulch can be expensive. Bark comes in various thicknesses and colors that are ideal for borders you want to cultivate. To reduce cost, you could put down a layer of cardboard or newspaper and then a thinner topcoat of bark. As organic mulches degrade back into the soil, they improve water retention and fertility.

There are many other organic mulches. Straw, for instance, can be used on strawberries to protect against heat loss and insects. Sand is perfect for protecting the crowns of dormant plants from frost and pests. After grass clipping mulch has disappeared over winter, the ground is littered with countless worm casts and the soil texture looks friable; a thin layer is best and it can become a bit like a layer of cardboard as it dries out in the sunlight. Try a small layer of grass clippings under soft fruit trees and among the vegetables. Shredded newspaper can be added as a mulch in the soil of a kitchen garden before planting potatoes or other vegetables. Pine needles are great as a weed suppressant and for retaining moisture.

Maybe the best organic mulch of all is simply plants! Yes, choosing ground-hugging plants around taller flora helps to retain moisture, and if you choose wisely, they may not greedily drink up the soil's moisture. Geraniums and nasturtiums are good ground cover and add valuable color. A tough plant, such as Lady's mantle, is good at suppressing weeds, holding onto moisture even in its leaves. Lemon balm is an excellent ground hugger and ornamental, too, with the variegated form being more attractive.

Nonorganic Mulches

Sometimes you just want to cover an area that you walk on between plants or maybe you just want to give an area that hot, arid look with dry stones. Whatever the color and type of pebbles, they can really complement your choice of planting. Grit and pea gravel do help to keep slugs down and retain water levels. Another big advantage is that these materials will remain, so you don't have to keep spending money and time replacing them. This convenience comes at a cost in the sense you've lost the value of the added fertility that organic mulch provides to the soil. Yet if you do have a dry, hot planting scheme that is truly drought-loving, this won't matter, as those plants prefer poorer soil.

The Advantages of Dry Gardens

It is actually less harmful to underwater a plant than to overwater it, especially in the autumn and winter when they are less active. Seasoned horticulturalists will even suggest underwatering a newly potted plant in order for it to stretch out its roots in search of water. This stimulates a good root system. When it comes to tomatoes, a dry spell may actually improve the flavor of the fruit.

Dry gardens can mean fewer pests, especially the dreaded slug or snail, which demands moist, damp conditions to survive. Dry gardens equate to fewer insects, because they need stagnant puddles to breed.

Avoid Frequent Waterings

To retain moisture in a houseplant while you are away, without drowning it, sit the plant on a sponge matting that holds a base of water in a tray. The plant slowly drinks up the water and can be left for days until you return.

To reduce the cost and burden of watering hanging baskets, mix moisture-retaining gel or granules into the soil. You can even purchase organic granules that break down into the soil over time. Both organic and inorganic types absorb and slowly release fluid into the soil when needed. Another solution is to use perlite—miniscule white pellets that aerate the soil and capture water to release when needed. Perlite is ideal in potted plants, as it helps to break up the soil and allows the roots to search out nutrients. It also prevents the soil from becoming water-logged and clogged, which can lead to rotting roots.

Capture and Store Precious Water

Beside using plants that need little water, there are plenty of ways to make the most of the water you do have.

Drain pipes to water butts: When the rain does fall, capture it. It makes sense to store rainfall for a non-rainy day. Attach a pipe from your gutter to a closed rain barrel and let it fill up for free. I station mine near where I water pots and baskets. You can always hide the water barrel by screening it with a climber over a trellis.

Tanks and barrels: Tanks and barrels can be left open to catch rainwater, which can be recycled. However, tanks and barrels should always be covered with a grill to avoid young kids falling into them. The water should always be used quickly; if it is left to go stagnant it becomes a breeding ground for bugs.

Irrigation systems and gullies: Dig a trench alongside a row of crops; the water in the trench or gully will slowly seep into the soil and water the plants around it. Instant irrigation system! You can let rainwater provide the water, or you can fill the gully yourself occasionally.

Recycling "gray" water: Dishwater or mild soapy water used for hand washing can be recycled into the soil and plant pots. The pots near my kitchen get "gray" watered, and they seem to like it. The mild soapiness can also kill off aphids.

Two Dry Gardens

The first dry garden design here is for a Mediterranean border, which suits the hot arid climate found in Northern Europe. Lots of plants from this region have gray or silver foliage as protection against hot summers, so they need dry, warm, sheltered gardens. These plants will do well in patios and terraces and where solid stone paths and walls help to retain heat in winter. Add a water feature or an eye-catching ornament to liven up the garden. One of the most luminous and visually soothing themes is a silver-and-gray garden with only white-flowering blooms. Although we automatically think of lavender as a blue floret, some varieties come in white. Snow-in-summer is ideal for flowing over stonework or as a ground hugger. Oriental poppy 'Perry's white' is also an eye-catcher. Other plants, such as lavender cotton, lamb's ear, cobweb houseleek, and members of the genera *Helianthemum*,

Artemisia (especially *A. abrotanum*), and *Dianthus* all do well under these sizzling conditions.

Silver-and-gray foliage plants may survive even in hot areas, depending upon the local climate. Some areas—such as the eastern United States, where summer humidity is high—make silver plants suffer because their leaf hairs hold too much moisture. Combat this by allowing plenty of air circulation, which reduces leaf rot.

Another design of a dry garden is the dry humid garden. There is a wealth of choice for the dry humid garden, from evergreen perennials, such as Achilleas; many Euphorbias, such as *E. nicaeensis*; to *Phlomis russeliana* or *Santolina pinnata*. Herbaceous perennials include acanthus, *Cynara cardunculus*, irises, and linarias. *Pennisetum setaceum* gives a wonderful show of flowers while sedum 'Carmen' helps to form a carpet of color. Let's not forget poppies, from Californian to oriental that thrive in poor, hot conditions. Tough customers, such as yuccas like *Y. gloriosa* 'Spanish dagger,' are evergreen, sword-shaped plants giving a huge panicle of ivory flowers in autumn. Most grasses can take prolonged dry spells, and some actually like dry conditions, such as Festuca grasses—its blue spiky foliage becomes even more vibrant in semi-drought conditions.

. . . .

Dry gardening involves ingenuity and thrift. Being economical with water can still open up a world of color and a garden that captures the eye. Once the groundwork is done and the plants are acclimated to their hot conditions, you can put your feet up, sit back, and bask in a relatively low-maintenance garden!

About the Author

Janice Sharkey is a keen Moon gardener. Having been a garden designer, she now enjoys nurturing her own garden and writing plays and children's fiction.

Cosmic Responses of Wildlife

By Bruce Scofield

Wild animals, like all other life forms on our planet, respond in many ways to the natural rhythms of the Earth and sky. The rotation of Earth, which causes the alterations of day and night, is the most important environmental signal that animals have adapted to. Activity and sleep cycles, called circadian rhythms, are the most basic behaviors set by this daily rotation. Circadian rhythms also drive chemical levels in animals' bodies that regulate hormones, metabolism, and other bodily cycles.

As the Earth orbits the Sun, the relative length of day and night shifts, a phenomena more pronounced as one moves away from the equator. Wildlife sense this changing light signal and many use it to regulate their sexual cycles, i.e., females coming into periods of fertility at the right time of year in order to enhance the survivability of any potential offspring. This is called photoperiodism. Animals also use the changing day and night ratios to initiate seasonal migrations to ensure a good food supply. Without these two primary solar signals, animals would have a hard time surviving.

The Moon is another signal that animals use to regulate their lives. Lunar signals are more complex than those of the Sun. We see the Moon rise and set, but not at the same time each day because the Moon advances in the zodiac and thus rises or sets about an hour later each day. The lunar or tidal day, from moonrise to moonrise, is 24.8 hours. Another lunar period is the synodic cycle, or the cycle from New Moon to New Moon, of about 29.5 days. Half of this, 14.77 days, which spans the period from New to Full Moon, is called the semilunar period. Marine animals are known to respond to some or even all of these lunar cycles, though how they do so is still a mystery. Although the Moon's cycles are not consistent with the steady beat of the solar year, the cycles are used by animals as triggers for behaviors like mating and migration and also to time gestation periods. The Moon's influence on the environment is not limited to variations in nocturnal light. The atmosphere itself is pulled by the gravity of the Moon and exhibits tides that modulate air pressure.

There are also solar-driven cycles that are longer than a year. The sunspot cycle has a powerful influence on many organisms, especially plants and insects. The two major solar cycles are the Schwabe cycle of 11.1 years and the Hale cycle of 22 years. The number of sunspots rises and falls over the course of the Schwabe cycle and when sunspots are plentiful, solar radiation is most intense. Every 11 years, the magnetic field on the Sun flips and this flipping is the basis of the Hale cycle. Since the amount of solar radiation affects many living things, including food for many animals, there are connections between these long cycles and animal populations and behaviors.

The Change of Seasons

The seasons change because day-length is modulated over the course of the year by the tilt of the Earth's axis. At the equator, day and night remain about the same length all year, but this changes in the higher latitudes. Beyond the Arctic Circle, day and night length varies in the extreme. This annual cycle of day-length in middle and higher latitudes has challenged the ability of animals to adapt, so they have evolved photoperiodic internal timing mechanisms that allow them to adjust to seasonal opportunities for feeding, reproduction, growth, molt, migration, and hibernation. It appears that the longevity of an animal is connected to the way it tracks the seasons.

Laboratory experiments have shown that many long-lived species have a free-running (environment-independent) biological clock that allows them to keep track of seasonal changes for many years without cues from the light signals of changing seasons. These are found in sheep, deer, bats, and starlings (a long-living bird), to name a few. Other shorter-lived species such as mice, hamsters, and many birds and reptiles have a seasonal rhythm that is more closely synchronized with the annual light changes. Changing day length also provides information on distance and direction that are crucial to migratory behaviors and, in mid to high latitudes, timing information for animals that hibernate for the winter.

Tidal and Lunar Rhythms

Lunar cycles are prominent features of many marine organisms. Some of the most studied tidal organisms are the several species of the fiddler crab (*Uca*). These arthropods live in burrows and feed as the tide ebbs and so they are subjected to changes of light, temperature, and tide. It has been found that they have a 24-hour circadian cycle of shell color—darker at dawn, lighter at dusk. However, their running activity shows a tidal periodicity of approximately ($\sim$) 24.8 hours, 50 minutes later each day. Further, the reproductive rhythm is semilunar (14 days), as the crab's larvae are released at New or Full Moon at the hour of high tide. It appears that more than one clock is operating in this organism.

An outstanding example of a response to the lunar synodic cycle is the reproductive behavior of the Palolo worms of the South Pacific. The end section of the body segments of this marine worm contains their sexual parts. These segments are released precisely at dawn with the October or November Full Moon. The Palolo worms require not only precise daily timing, but they also must time this sexual event relative to the seasons. Records of Palolo spawning events show that, by following the Full Moon in either October or November, the worms accurately track the 18-year Metonic cycle, wherein a specific phase of the Moon occurs on the same date every eighteen years. The same sort of timing is found in the reproductive actions of at least 107 species of coral along 500 kilometers of the Great Barrier Reef in the western Pacific. Nearly all the coral release their reproductive gametes three to six days following the Full Moon in October or November and about four hours after sunset.

At present it is not known for certain how lunar signals are internalized in animals. Light is certainly a factor in many organisms, though the Moon's light is minuscule compared to that of the Sun. Fish appear to follow Moon cycles, and it seems they respond to the varying degree of lunar light through the course of the twenty-nine-day synodic cycle from New to Full Moon. Studies of juvenile salmon show that they group together at Full Moon and begin to move or migrate at New Moon. It is thought that this strategy limits predation, but there may be other reasons as well. Young fish feed on zooplankton, and these tiny crustaceans apparently move from shallow to deep water in response to moonlight levels—a vertical migration. If there is too much light (at Full Moon), zooplankton will migrate to deeper water to avoid predators, and the juvenile salmon (and probably many other fish species) will follow them down to depths where little light penetrates. So the phase of the Moon regulates behavior of the zooplankton, and the fish that feed on these smaller organisms will then entrain to the lunar cycle as well.

A deeper kind of lunar rhythm may be found in marine organisms called foraminifera. These critters, which are not animals but members of a kingdom of life called Rhizaria, have a complicated reproductive cycle. Reproduction includes an event wherein hundreds of thousands of free-swimming reproductive cells—gametes—leave the mother foraminifera and meet with others of its kind to fuse and begin the next phase of the life cycle. Reproduction in the ocean requires a large number of gametes, about 300,000–400,000 in order to avoid dilution, and it requires consolidation in time and space. To get this right, the foraminifera has to synchronize its gamete release precisely along with others of its kind, and the Moon is apparently the signal. A study found that each foram species responds to a specific day in the lunar cycle most near to the Full Moon.

The Geomagnetic Field

The Earth has its own self-generated magnetic field called the magnetosphere. The magnetosphere has fluctuations and periodicities that range from less than 1 second to longer ones that include responses to the Earth's rotation (24 hours), the lunar day (24.8 hours), the synodic month (29.5 days), solar rotation (~27 days), the solar year (365 days), variations in the solar wind (1.3 years), and the sunspot cycle (11 and 22 years, on average). Clearly, the magnetic field is a potent information source for life on our planet. The Earth's magnetic field is utilized by certain organisms for location in space and time, valuable information for activities like feeding and migration. To feed or migrate, an animal must navigate, and this is the most studied use of the magnetic field by organisms. The ability to "read" the geomagnetic field requires remarkable discrimination, but it is accomplished by many animals—and also by bacteria that use it to migrate up and down in the water.

Magnetotactic bacteria are aquatic, anaerobic (oxygen-avoiding) microbes. They move in the water by means of flagella and navigate by aligning themselves with the Earth's magnetic field. At middle latitudes, the magnetic field is more vertical than horizontal as it is plunging toward the poles. To avoid oxygen, magnetotactic bacteria use the north-sloping direction of the Earth's magnetic field as a kind of vertical guide in order to find places with very low oxygen levels. Their response and alignment to the magnetic field is possible due to a chain or chains of small, biologically produced crystals of magnetite inside the body of the bacteria. These particles are called magnetosomes.

Magnetosomes are single crystals of magnetite (Fe_3O_4) synthesized by the organism from materials in their environment. Magnetite is the mineral that forms lodestone and is known to form inorganically only at high temperatures and pressures. Other organisms known for their navigating abilities have also been found to contain magnetite crystals in their bodies. This list includes pigeons, which have magnetic particles in their skulls. These birds are able to navigate in three different ways: by using the Sun, visual cues from topography, or the magnetic field—quite a backup system, indeed! Honeybees have magnetite in their abdomens that they use to locate food sources relative to the hive. Other animals with magnetite include tuna, trout, blue marlins, green turtles, whales, and dolphins, which migrate through the oceans at depth, without light cues. It is possible that humans have magnetite particles in the brain to aid orientation. One study tested a group of young students by taking them blindfolded on a bus to a distant location and then asking them to find the direction from which they came. This study suggested that people do sense direction, but other studies haven't produced similar results. It has been suggested that magnetic sensory systems evolved early in life history and are completely separate from other sensory mechanisms.

Solar Cycles

Sunspots are visible evidence of both solar rotation and of a larger cycle of solar magnetism that affects the Earth's magnetic field. The numbers and groups of sunspots changes daily, and their positions on the Sun's surface changes over time. The most recognized solar cycle is the Schwabe cycle of ~11 years, during which the number of Sunspots increases and decreases. The Hale cycle of ~22 years accounts for the magnetic reversals that occur between Schwabe cycles.

Many reported correlations link wildlife and solar activity. Well-documented rhythms of about ten years, close to the Schwabe cycle, include crop yields and fish catches. Insect populations show very close correlations with the Schwabe cycle and, as such, are regarded as sensitive climate monitors. Tent caterpillar populations peak predictably about two years before the peak of solar activity, and other insect populations appear to do the same. It has been suggested that these population increases are due to the increased warmth and ultraviolet radiation that also follow the solar cycle. Like the juvenile salmon that follow the zooplankton, other animals will likewise follow this cycle in population and possibly migration patterns as they depend on insects for food.

In the boreal forests of North America, fur-trading records since 1671 reveal that the snowshoe hare's population fluctuates in roughly ten-year cycles. Exactly why is not clear, but the population of the hare's predator, the lynx, also follows this rhythm, which is suspiciously close to the Schwabe sunspot cycle. Other animals seem to follow this pattern as well. Porcupine abundance has fluctuated since 1868 in eastern Quebec, and a correlation has been found between the porcupine population cycle, fluctuations in local precipitation and temperature, and the solar cycle. Winter precipitation plays a key role in porcupine population dynamics. In porcupine populations, both the 11- and 22-year periodicities are found, though only the 22-year cycle correlates

with snowfall. Further, the link between the solar cycle and climate appears to be localized to the study region, suggesting that
porcupine abundance cycles may exist only in locations with
strong climatic cycles.

. . . .

The links between wildlife and the cycles of the Sun and Moon are
quite strong, but the actual connections between sky and Earth
are apparently not consistent. In some species, light is the key;
in others, the changing tides; and in still others, weather cycles
that are driven by solar cycles affect prey and, in turn, predator.
One thing for certain is that life on this planet is connected to the
movements of astronomical bodies. Sounds like astrology to me.

About the Author

*Bruce Scofield is a practicing astrologer who has maintained private
practice as an astrological consultant and conference speaker for
more than forty years. He is the author of seven books and more
than two hundred articles on astrology. He has served on the education committee of the National Council for Geocosmic Research since
1979. He holds a master's degree in history and teaches astrology
and psychology at Kepler College and Gaia theory and evolution at
the University of Massachusetts. Presently, he is working on a PhD
program in geoscience (solar system influences on climate and life)
at Umass. Scofield and Barry Orr maintain a Web site (www.oner-
eed.com) with articles and a calculation program on Mesoamerican
(Maya and Aztec) astrology.*

References

Brown, Frank A., J. Woodland Hastings, and John D. Palmer. *The Biological Clock: Two Views.* New York: Academic Press, 1970.

Dunlap, Jay C., Jennifer J. Loros, and Patricia J. DeCoursey. *Chronobiology:
Biological Timekeeping.* Sunderland, MA: Sinauer Associates, 2003.

Endres, Klaus-Peter, and Wolfgang Schad. *Moon Rhythms in Nature.* Ed-
inburgh, UK: Floris Books, 2002.

Taylor, Bernie. *Biological Time.* Oregon: The Ea Press, 2004.

Ward, Ritchie R. *The Living Clocks.* New York: Alfred A. Knopf, 1971.

2009 © Rob Marmion. Image from BigStockPhoto.com

Job Hunting by the Moon

by April Elliott Kent

"I just had another horrible job interview," my friend Brenda groaned. A casualty of the mortgage industry meltdown, Brenda—a bright, capable woman in her late forties—lost her job to downsizing for about the fourth time in as many years. "I can't stand any more interviews!" she declared. "I've got years of experience. I'm great at what I do! Why can't I get a job?"

I listened to Brenda's plight with sympathy, but also with a guilty sense of relief. I'm self-employed, and it's been years since I interviewed for a job. It's a ritual I don't miss, especially in today's softening economic climate. And really, does anyone ever enjoy it? A job interview is about as much fun as a blind date: you have to look your best, quell your rising levels of desperation and anxiety, and behave charmingly under pressure. And just as any blind date is a gamble, there is no guarantee that you'll "click" with your interviewer—provided you land an interview in the first place.

The self-employed aren't completely off the hook, either. After all, the process of wooing prospective customers and clients can feel remarkably like a job interview. If you're a contractor or business owner, it can be challenging to gain an advantage over your competition and win the trust and business of customers.

Assuming you've got a fantastic resume, have done your research, and don't walk into appointments with grime under your fingernails, how can you stand out from the competition? Basic astrology—specifically, the sign and aspects of the Moon—can help you make a strong impression and improve your chances of landing a great job.

The Mood of the Day

As anyone in advertising can tell you, when you're preparing a pitch—whether for a job, a client, or a prospective romantic partner—it's essential to understand the mood of the room. Larger planetary trends on any given day are certainly important, but the Moon, as the astrological ruler of daily affairs, describes the atmosphere on the ground. How is everyone feeling? To which things, people, colors, and qualities are we instinctively drawn? Are people especially impatient, cutting each other off in traffic and exchanging rude gestures? Chances are good the Moon is in Aries or in aspect to Mars. Overwhelmed with a sudden desire to clean out your filing cabinets? Hello, Moon in Virgo!

Understanding the Moon's sign and aspects on a given day can help you better tailor your presentation to your audience. And depending on your own Sun, Moon, or rising sign, you gain the upper hand on days when the Moon is in certain signs, as others are more receptive to your natural style during those days.

Lunar Aspects

The aspects formed by the Moon to other planets are another indication of the day's mood. Look at the table of aspects included beginning on page 140 of this book to find the Moon's aspects on the day of your interview. A guide such as *Llewellyn's Astrological Pocket Planner*, which lists the exact times of these aspects, can also be useful; the lunar aspect closest to your appointment will nearly always be quite descriptive of the interview.

A Note About the Void-of-Course Moon

If you have the option, avoid scheduling interviews for times when the Moon is void-of-course (see tables beginning on page 80). These interviews are more likely to be cancelled or rescheduled, and if they do take place, they may be less apt to result in a job offer. Often, the void-of-course Moon signifies that the position itself will be cancelled before it is filled or that a candidate has already been essentially chosen. This is less likely to be the case if it's your second or third interview for the same job. In these cases, the void-of-course Moon may indicate that the candidate who has already been selected is you!

Twelve Moons, Twelve Moods

Moon in Aries (or approaching aspect to Mars):

Today's mood is colored by impatience, so be prepared to get to the point and don't be dismayed if your interviewer is terse or hurries through your interview. Prepare concise answers to predictable questions, speak quickly, and be assertive. It's an especially good day to interview for jobs in sales, mechanics, athletics, or those connected with the military. *Advantage to Aries, Leo, Sagittarius and to some extent Gemini, Scorpio, and Aquarius.*

What to wear: Aries, a fire sign, favors all shades of red to make you seem more dynamic and confident. Aries rules the head, so pay special attention to your hair and teeth. But don't go overboard: you should look tidy and attractive, but streamlined and ready for action. Aries is attracted to whatever is new and trendy, so don't be afraid to break out an outfit or accessory that pushes the envelope of professionalism just a little.

Moon in Taurus (or approaching aspect to Venus):

The hectic pace of the Moon in Aries gives way to a more relaxed mood today, and good manners count. Taurus is a sign that respects confidence and competence but has little patience for showboating or gross displays of egotism. Present your case thoroughly, emphasizing practical skills and common sense, but don't oversell. It's a good day for interviews related to work in finance, agriculture/gardening, and engineering. *Advantage to Taurus, Virgo, Capricorn and to some extent Cancer, Libra, and Pisces.*

What to wear: Calming, earthy Taurus has an eye for luxury, so pull out your most classic and expensive outfit in the richest textures and fabrics. Wearing soft shades of rose and blue will help you project a soothing image, and a hint of very subtle fragrance can be appealing.

Moon in Gemini (or approaching aspect to Mercury):

Conversation rules the day, so be prepared for your interview to begin and finish late to accommodate a lot of chitchat. It's more important today to sell yourself as a person who's fun to have around than it is to offer an exhaustive recitation of your achievements and qualifications. It's an good day to interview for jobs in communication, media, sales, and advertising. *Advantage to Gemini, Libra, Aquarius and to some extent Aries, Leo, and Virgo.*

What to wear: We seek variety when the Moon is in Gemini, so contrasting colors and patterns, if chosen carefully, can be effective today. An interesting accessory, such as a lapel pin or cufflinks, demonstrates attention to detail and appeals to Gemini's curiosity.

Moon in Cancer (ruled by the Moon):

Emotions run high today, and the outcome of any interview is unpredictable. Even if you make a great impression, by the time the Moon enters Leo, your interviewer may have decided to go with a completely different type of candidate. Be calming, steady, and sensitive to the mood of the room—but not oversensitive. It's a good day to interview for jobs in public relations, health care, and food service industries. *Advantage to Cancer, Scorpio, Pisces and to some extent Taurus and Virgo.*

What to wear: Conservative clothes in shades of soft gray, silver, or blue are the order of the day. Cancer has a nostalgic streak, so old-fashioned touches such as a pocket watch or pearl earrings are a respectful nod to tradition.

Moon in Leo (or approaching aspect to the Sun):

Everyone wants to feel like a star today, so convey genuine admiration for your interviewer and the company that he or she represents. It's always a good idea to send a thank-you note after an interview, but even more important when the Moon is in Leo. This is a good day to interview for jobs requiring performance skills (teacher, trial lawyer, actor, trainer), working with children, or working in arts or entertainment (casinos, theaters, theme parks). *Advantage to Aries, Leo, Sagittarius and to some extent Gemini and Libra.*

What to wear: People expect some color and razzle-dazzle while the Moon is in dramatic Leo. Colors can be bold—scarlet, gold, orange—but the cut and fabrics should be classic and impeccable.

Moon in Virgo (or approaching aspect to Mercury):

Today, project an image of competence, precision, and attention to detail. Have your resume double-checked by a professional to make sure it is perfect, and ensure that your grooming is flawless—you will definitely be judged for your missing buttons and rundown loafers. It's a great day to interview for jobs requiring organization, logic, and problem-solving ability, such as administrative assistant, health care professional, accountant, information technician or programmer, or administrator. *Advantage to Taurus,*

Virgo, Capricorn and to some extent Gemini, Cancer, and Scorpio.

What to wear: Classic gray and navy blue are safe bets for Virgo Moon days, with perhaps a bit of pinstripe or an interesting weave to provide some texture. Wear simple jewelry and accessories, and practical, well-made shoes.

Moon in Libra (or approaching aspect to Venus):

Today of all days, interviewers want to see a team player. Emphasize your ability to get along with others, to see both sides of an issue, and to put the customer first. It's a great day to interview for jobs that emphasize public relations, sales, customer service, advocacy, and human resources. *Advantage to Gemini, Libra, Aquarius and to some extent Taurus, Leo, and Sagittarius.*

What to wear: Libra is especially attuned to fashion, so wear fine fabrics such as silk or linen in soft shades, perhaps with accents of soft rose or blue. Your makeup, hair, and nails must be impeccable; your accessories, tasteful but elegant. A faint whiff of fragrance is fine, but keep it very light—nothing heavy or musky.

Moon in Scorpio (or approaching aspect to Mars or Pluto):

The Moon's passage through Scorpio is a time of great creativity, and most interviewers would rather be doing their own work than interviewing you. Don't attempt to charm or cajole your way into the job, just be honest and direct about your experience and qualifications. Good eye contact and a firm handshake are vital. It's a good interview day for jobs related to security, law enforcement, or detective work; investments, insurance, or mortgages; or extreme environments or circumstances, such as emergency response, toxic substances, or death and dying. *Advantage to Cancer, Scorpio, Pisces and to some extent Aries, Virgo, and Capricorn.*

What to wear: Scorpio favors strong colors, such as crimson or even a touch of chartreuse; use them to accent a strong black or charcoal suit with a strong cut. Women can get away with wearing a slightly stronger shade of lipstick than usual, along with a bit of smoky eyeliner and shadow—but don't overdo it.

Moon in Sagittarius (or approaching aspect to Jupiter):

The brooding Scorpio Moon gives way to a brighter, cheerier mood as the Moon enters Sagittarius. An upbeat, positive attitude and a quickness to laugh will endear you to interviewers. Sagittarius is also the sign of higher education, so credentials count. It's a good day to interview for jobs in academia, the travel industry, and intercultural affairs. Sagittarius' affinity for large animals also makes it a good day to interview for jobs working in zoos, for veterinary practices, or at racetracks. *Advantage to Aries, Leo, Sagittarius and to some extent Libra, Aquarius, and Pisces.*

What to wear: When the Moon is in Sagittarius, people respond well to deep blues, teal, and touches of plum. Wearing a necktie or scarf with an exotic pattern or an accessory from another culture will excite the Sagittarian love of foreign lands.

Moon in Capricorn (or approaching aspect to Saturn):

People feel overburdened today, and your interviewer probably sees you as another unwelcome task to be crossed off his or her "to do" list! Present yourself as serious, responsible, respectful, and extremely hard working. Address interviewers using last names unless invited to use first names. Today is a good day to interview for management positions and for work in industries that are conservative and traditional (engineering, financial services) or related to physical labor (especially construction, concrete, and masonry). *Advantage to Taurus, Virgo, Capricorn and to some extent Scorpio, Aquarius, and Pisces.*

What to wear: The most traditional of signs, Capricorn responds to conventional clothes in classic shades of black, white, blue, and gray. Don't skimp on quality or cut—well-tailored suits and good-quality shoes lend an air of prosperity, dignity, and tradition that appeals to Capricorn.

Moon in Aquarius (or approaching aspect to Saturn or Uranus): Aquarius is the sign of rule-breakers, so you can get away with displaying a spark of individuality in most interviews. Be yourself, but demonstrate that you can find common ground with absolutely anyone. Treat everyone you meet, from janitors and receptionists to the CEO, with warmth and friendly respect. It's a good day to interview for jobs in public relations, politics, start-up companies, technology, and media. *Advantage to Gemini, Libra, Aquarius and to some extent Aries, Sagittarius, and Capricorn.*

What to wear: Bright shades of blue are Aquarian favorites, along with flashes of silver jewelry. Underneath its friendly exterior, though, Aquarius has a strong conservative streak. The rule of thumb is to match one of your expensive, conservative Capricorn suits with one shiny, unexpected accent color or accessory.

Moon in Pisces (or approaching aspect to Jupiter or Neptune): The mood is either empathetic or self-pitying today, and interviewers are inclined to be a little more talkative—and sometimes a little indiscreet. Flow with the mood of the room as closely as possible without expressing any negativity. It's a good day to interview for work in health care, social work, the arts, metaphysics, travel, or intercultural environments.

What to wear: Expressive, artistic Pisces is perhaps the trendiest of all signs, so pull out one of your more fashionable outfits. Pisces favors the colors of the sea—soft greens and blues, turquoise—and flowing, softly draped fabrics. *Advantage to Cancer, Scorpio, Pisces and to some extent Taurus, Sagittarius, and Capricorn.*

Getting in Sync

When economic factors are against you or if your qualifications are not all they could be, you'll need a lot more than the Moon's help to improve your odds of landing a job. Brenda, who's looking for work in a hard-hit sector of the economy, is still interviewing. Since she began following the Moon, though, she claims that at least her interview experiences have been more pleasant.

In the end, of course, not even a thriving economy, years of preparation, and great qualifications can guarantee that you'll land your ideal job. You can be the best-qualified applicant for a position yet fail your interview simply because you didn't click with the personality of a particular interviewer on a given day. Knowing and getting in sync with the mood of the day can help you appeal to interviewers on a critical, subliminal level, and ensure they are left with a good feeling about you, even if they're not sure why. So, polish your resume, buff your best shoes, dry-clean your best outfit—and make the Moon your employment agent!

About the Author

April Elliott Kent, a professional astrologer since 1990, is a member of NCGR and ISAR and graduated from San Diego State University with a degree in communication. Her Star Guide to Weddings *was published in 2008 (Llewellyn). April's writing has also appeared in* The Mountain Astrologer *(USA) magazine, the online journals* MoonCircles *and* Beliefnet, *and Llewellyn's Moon Sign Book (2005, 2006, 2007, 2008, and 2009). April lives with her husband and two cats in San Diego. Her Web site is: http://www.bigskyastrology.com/*

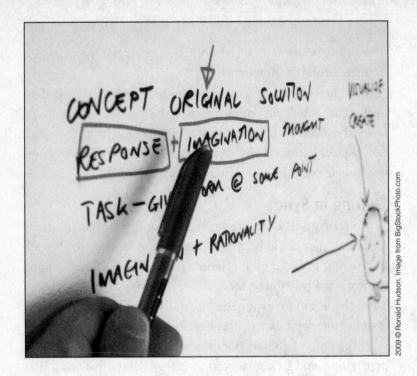

Grow Your Business by the Light of the Moon

by Dallas Jennifer Cobb

We all need to work, to make money, to feel vital, and to contribute to our communities. I worked for many years in a full-time occupation that was engaging, satisfying, and financially rewarding. I enjoyed work and literally lived to work.

But, as it is apt to do, life changed. A car accident left me with residual injuries that affected my ability to do my job. Not long afterward, I became pregnant and soon had a baby.

As I was trying to balance the needs of my delicate health and the needs of a growing baby, full-time work no longer fit my life. My priorities had changed, and I didn't "live to work" anymore. I started thinking about simply "working to live," having work that fit around my life and allowed me more time with my child and to take care of myself. This was a big shift from organizing everything in my life around my work!

Because it is hard to find a job that fits around life, I ended up creating my own job. I planned, started, developed, and "grew" my own business, creating satisfying work that fit my life and the lives of my family. It's been seven years since my child was born, and I am still happily self-employed in a variety of realms. I enjoy an abundance of time with my family, money, and satisfying work. Providing you with some relevant tools and resources might just inspire and motivate you to "grow" your own business!

Lunar Tips

In general, the increasing energy and size of a waxing Moon is the best time to initiate new things, make plans that you want to come to fruition, and draw things toward you. The Full Moon is a time of great clarity because of the increased light and lunar energy. It is the time of the Moon's greatest power, a time for acknowledging your abundance, taking stock of your blessings, and seeing yourself and your business in the best light. The declining energy of a waning Moon is the time for fine-tuning and evaluation, ridding yourself of unnecessary or outdated habits, trimming excesses, and getting rid of unwanted people or energy. And the New Moon is a time for evaluation, reflection, and adjustment of your business practices, preparing to launch into new patterns or habits with the light of the New Moon.

The Moon is continuously moving through the twelve signs of the zodiac. Each of the zodiac signs has its own particular influence, so I have included suggestions for their use.

For example, if you want to do work that grows your business, but the New Moon phase is still a long way off, you could choose to do this work when the Moon is in Taurus—an energy that lends itself to longevity, growth, and staying power.

For more Moon phase and sign information, consult the almanac section of the current *Llewellyn's Moon Sign Book*.

A Job?

Although I had income while on maternity leave, I started exploring my options. I considered part-time work, but realized that the cost of child care almost negated any earnings.

Then I looked at what work would really cost me. Not just child care, but commuting costs, additional vehicle insurance, meals while at work, and the added cost of maintaining my vehicle given the extra wear and tear. There was the cost of a professional wardrobe, hair care, and cosmetics. But more than just monetary costs, I had to consider the cost to my health from increased stress and the long-term emotional cost of having someone else raise my daughter.

An accountant helped me analyze our household income, expenses, and taxes. My partner had a fixed professional income, which put us in a high tax bracket and allowed no significant write-offs. After calculating the cost of working and the requisite income taxes, we determined I would have to make in excess of $60,000 per year before we would see a net gain in our household income. Convinced I didn't want to work full-time, my partner and I concluded that we were further ahead if I stayed home.

But I still wanted meaningful work, and I wanted to contribute financially to our household, so I began to research self-employment. In Canada, our tax laws allow self-employed people to write off expenses against their income. A self-employed person can write off not just office supplies, but a percentage of housing costs, utilities, vehicle costs, office expenses, education, business-related travel, even child care. From a financial point of view, a small business looked beneficial to both me and my partner.

Why not operate a small business, enjoy flexible hours, make some money, and take tax write-offs against both my income and my partner's? It sounded ideal.

Lunar Tips

Cull unwanted things with the waning Moon, when the decreasing energy will contribute to culling and letting go. With a Moon in Aquarius, draw on the energy of the rebel to break old habits and make changes that benefit individuality and freedom. A Scorpio waning Moon can help retard growth in one area, while focusing energy on fruition in another.

The best time to do financial planning is with the New Moon. It is a time to initiate new ventures. The first day of the New Moon is the day to undertake rituals to increase money. Do your money planning with a Moon in Taurus for an increase in value and solid longevity.

A Business?

There's lots of information available about starting and running a small business. The literature I was drawn to focused on the stories of women like me, whose prime motivation was to stay at home with their kids. Identifying themselves as work at home moms (WAHMs), their stories detailed not just their business experience, but how they balanced the business with their family. Their stories spoke of flexible hours, home-based activities, the opportunity to have their children with them, and the fulfillment of both making money and fully parenting their children.

This arrangement sounded right to me. Yes, I wanted to run a business instead of working in a nine-to-five job, but what kind of business would I run?

Like trying to decide what to eat in a restaurant with a large menu, trying to decide what kind of business to run was mind-boggling. The possibilities were endless. Uncertain what to choose, I refocused on the things that motivated me initially. I wanted work that fit my life, which meant flexible hours, a small financial investment, and something enjoyable.

I started to think about what I love doing and let that guide me.

Lunar Tips

Research is best done during the waxing phases of the Moon. As the Moon grows, let your knowledge of a subject grow, shedding light on new paths, skills, and venues. Reflecting on information can then be done with the waning Moon, as it is a time for introspective study, meditation, and creative magic. A Moon in Capricorn provides strong structures to support business growth.

How Life Shapes Business

Designing a business to fit my life sounded good, but what did that really mean? At first I focused on how I would like to work, hoping that the what of my business would come. Ask yourself a few questions:

Do you prefer to work alone or with others? Would you rather work with people or with things? Do you like to do one thing repeatedly or manage a variety of tasks? Do you have any skill or ability that you think you could sell? Is there some focused training or education that could certify you in something saleable?

I enjoy working with others, and I prefer to work with people. I prefer a variety of tasks, so I don't get bored. I have a background in adult education and lots of teaching experience. And, if I took some training that would certify me to teach fitness classes, it could unify my love of fitness with my skills in adult education.

Lunar Tips

Moon in Leo draws emphasis to self, for a focus on your own needs and wants. Full Moon energy helps you to brainstorm and make intuitive leaps between ideas.

Do What You Love

I read *Do What You Love, the Money Will Follow* by Marsha Sinetar (Dell Publishing, 1987). While I hate empty, glib statements, the truth of this has worked for me. Because I wanted to enjoy my work, I decided to look at what I loved in my life. Could I turn any of those activities into a business? I wrote a list: I love my daughter; I love books, reading, and writing; I love Pilates, running, and working out.

I started to think more about Pilates, an exercise modality that focused on increased strength, flexibility, and mobility. I was paying to take Pilates classes in a neighboring community. What if I learned how to teach Pilates? I could be paid to do something that I am currently paying for, so I'd be making money and saving money at the same time. Plus, I could offer Pilates classes closer to home, tapping into the potential local market.

Lunar Tips

Brainstorming is an activity best done just before or on the Full Moon, when the energy of the Moon is abundant, brilliant, and inspiring. The Full Moon is also especially potent for matters of the heart, and finding your right work is definitely a matter of the heart. A Moon in Pisces will put you in touch with dreamy energy, intuition, and psychic influences.

Low-risk Business Start-up

Because the transition to self-employment is a big one, it is useful to create a safety net for support during the transition. I had a variety of safety nets.

While on maternity leave, I had income support. I used it to fund my Pilates teacher training. Once certified, I applied to and was accepted into a small-business training program that provided me with ongoing income support and business coaching during my business' start-up phase.

Keeping the risk factor low also means keeping your start-up costs low and your income-to-expense ratio high. That means you must start your business without spending a lot of money. I needed training for certification, a computer to do bookkeeping and create advertising materials, office supplies, a portable CD player, music CDs, Pilates/yoga mats, and a car to get around in.

Most of these I owned already and could use in my business. So all I purchased was six more Pilates/yoga mats, some office supplies, and a couple of CDs. With a minimal capital outlay, I was ready to jump into business and quickly generate profit.

Lunar Tips

Moon in Libra energy will promote balance, partnership, and cooperation—all energies that contribute to a strong safety net. Sagittarius also favors expansion, growth, and confidence.

Selling Yourself

The easiest way to sell something is to take the time to research what the buying public needs and wants, then meet that need. A niche market is a specific market with specific needs that (hopefully) are not currently being met by your competitors. Knowing your target community, or niche market, will enable you to slant your services and advertising to their needs.

Here in North America, there is an aging population, as the Baby Boomer generation hits retirement age. I saw this active, older population as a niche market: lots of people with lots of money who wanted support to stay active and healthy. To target this group, I designed classes, lesson plans, and marketing materials to appeal to their needs. With a focus on maintaining mobility and function and preventing injuries, my classes appeal to people fighting the effects of aging. Pilates also appeals to women, so I made sure my advertising appealed to them.

Lunar Tips

Moon in Cancer will tune you in to others' needs and will support growth and nurture of your business. When starting a new business, look for an increasing Moon in Taurus, Virgo, or Capricorn to tap into earthy energy, which will nourish your venture.

Writing a Functional Business Plan

A business plan is a written statement that describes you, your business, and where it is going. It is a document that you can use to apply for business startup support, financing, or investment. It specifically describes your business timelines, financial management, and development plans.

Not just for other people, a business plan is also a planning tool for your use. By drafting a two-year plan and putting it down on paper, you have a template to measure your development against, judge your continued success or failure, and track your progress.

Most commonly, a business plan has a cover that states the company name, owner's name, the business address, and contact information. It also has a table of contents.

The "Summary of the Business" section includes a business description and a management resume highlighting the skills, experience, and education that you bring to the business.

The "Organization of the Business" section includes information on your hours of operation, what employees or associates you'll need, and an identification of the professional services your business will require, such as a lawyer, bookkeeper, accountant, or insurance agent.

The "Marketing" section of the business plan identifies your target market and summarizes the market research you have done. This section also contains information on the benefits of your product or service, a discussion of the competition in your area, and your marketing and promotional plan—a timeline that details how and where you will market your business.

The "Financial" section of the plan summarizes your projected startup and operating expenses; your sources of financing; how you arrived at pricing and costing your product or service; and detailed information on cash flow (commonly done as a monthly cash flow forecast), summarizing expenses and income for the initial twenty-four-month period.

The "Conclusion" section details why your business will be successful and sets the tone for potential outside evaluation.

Lunar Tips

Moon in Capricorn will put your focus on structure as you write your business plan, and good structure provides a place for a creative business to grow. Started during the dark of the Moon, a business plan can involve deep searching within the self. As the New Moon arrives, continue to edit and rewrite, bringing the document into the light of being, growing your ideas with the Moon's development.

Marketing with No- or Low-cost Ideas

I didn't have much money when I started my business, and I didn't want to amass more debt, so I brainstormed ways to market my business with very little or no money. I told my friends, family, and former coworkers about my Pilates classes, inviting them to attend one. I designed and printed advertising posters that I put up for free in grocery stores, post offices, banks, restaurants, and laundromats in my community. I also designed and printed a small handbill that I gave to everyone I talked to. I organized a free introductory Pilates class and invited participants to then sign up for a series of classes, which they would pay for. I joined the local business association and had a new business profile printed in the local paper. I joined a small business network, where I met other small business owners to share ideas, collaborate, and promote our businesses.

Don't be afraid to try imaginative marketing schemes that are cheap or free. They can often payoff in increased business.

Lunar Tips

Launch marketing campaigns with the New Moon to harness the growing lunar influence. Moon in a sign such as Cancer, Scorpio, or Pisces promotes fluidity, growth, and fruition. Moon in Gemini brings enhanced communication, and Moon in Capricorn will help to build business.

Financial and Legal Business Planning

There are some aspects to business that cannot and should not be done on the cheap, such as finding good legal and financial advice. Research your potential liabilities and take steps to best protect yourself from them financially and legally.

When you need help, ask an expert. You may find one who is willing to barter information in exchange for your service or product. If not, think of these fees as money well invested in your future peace of mind.

Lunar Tips

Moon in Capricorn provides a productive, dry energy that is ideal for setting boundaries and rules, developing strong structure, and observing traditions and obligations.

Jumping In

There is an old saying in business: No one plans to fail, but lots of people fail to plan. With some of the information and tools provided, why not start to grow your business by the light of the Moon? Once the seed of imagination is planted, use the finite tools to give structure to your creative ideas, and—using the fertile energy of the Moon—let your dream work grow to fruition. Don't be satisfied just to live to work, but strive to work to live, and live a little while you're doing it.

About the Author

Life is what you make it, and Dallas Jennifer Cobb has made a magical life in a waterfront village on the shore of Lake Ontario. Forever scheming novel ways to pay the bills, she currently teaches Pilates, works in a library, and writes. A regular contributor to Llewellyn's almanacs, she wrote two novels this year with the support of National Novel Writing Month (NaNoWriMo.org). You can contact her at: jennifer.cobb@sympatico.ca

Judgment and The Moon

By Lesley Francis

It might seem a bit silly to say that each day begins with a judgment, but think about it. Even before your eyes really open —possibly unwillingly—you begin to size up your day. All your senses start to take in available information. Light, sound, smells, body sensations, etc. Is it sunny? Are the kids—whether of the human or animal kind—up and about? Has somebody made coffee? Does it feel chilly? And while you are processing that sensory input, your mind will probably jump to what's on the agenda for the day. "Okay, today is Tuesday. Today I have to _____."

So, by now, complex being that you are, you're ready to make your very first important judgment of the day (after all, you've already absorbed and judged a lot of things in the 120 seconds since you woke up).

Do you want to get out of bed or don't you?

At this point, you may find yourself in conflict . . . with yourself. It could be you just don't want to get out of bed. Or maybe there's something in your daily routine—like washing your hair—that you really hate (I'm with you on that one, especially the blow-drying). Or you have a long list of really boring things to do. Or you're itching to leave on holiday but have a deadline hanging over your head. So, you can see the possibilities for internal jibber-jabbering are endless.

You flip-flop back and forth, weighing a number of factors—practical, emotional, organizational, consequential—so you can get on with life. All this involves a number of judgments. "Is this important? Is that? What about X, Y, and Z?"

And you aren't even out of bed yet! Overwhelmed? Don't be. All this just demonstrates one really obvious truth: Life is a series of judgments. Each day you make hundreds of them.

Judgments range from the practical (Is it safe to cross the road?) to the organizational (Can I really fit all those things into one day?) to the consequential (If I paint my bedroom chartreuse, will I be able to live with it?) to the emotional (How can I tell my friend that I don't like it when she yells at me?).

You can't make a decision without making a judgment. It's all part of figuring out what's going on. So much so that you're not even aware you're doing it.

So embrace your capacity for judgment. Without it, how could you possibly survive and thrive? Just remember, judgment is a powerful tool. And, to make the best use of it, you need to know and understand that judgment is rooted in how you feel not how you think. How you feel about everything, from green beans to politics to housecleaning to the girl your son is currently dating.

Which leads us directly to the Moon. She colors everything we do—our responses, our habits, our inclinations, our motivations, our reactions, and, yes, our judgments. How we feel fuels everything in our lives. The key is to be tuned in to the fluctuations of our inner world and how they influence our perceptions and, ultimately, our judgments and our actions. Otherwise we remain at the mercy of those fluctuations, unable to make the most effective use of our experiences because we are temporarily blinded by the ups and downs of our emotional landscape.

How do we get tuned in? The answer is obvious: the lunar cycle! It is a powerful descriptive tool for understanding how life unfolds, and it is deeply imprinted on our consciousness. It operates whether or not we are aware of it, so why not use it in a conscious way to illuminate what we are already experiencing? In fact, the Moon calls upon us to recognize that she is not just pure feeling, she is also our guide to a healthier life, physically, mentally, emotionally, and spiritually. And each of her phases is a signpost on that journey. Step-by-step, Moon phases show us how to engage fully in the process of life and emerge renewed, refreshed, and ready for more. Nowhere is this more important than in the constant need to make judgments.

Before we explore the use of the Moon's cycle to expand our capacity for judgment in a positive, life-affirming way, there are a couple of things you need to keep in mind.

One is that you can use the lunar cycle as a symbolic template, you can use it quite literally, or you can do both. The first allows you to apply a known process to your daily life, especially when you are dealing with stress that is blocking your path; the second offers you a chance to go deeper, to really feel the cycle working by tuning into its natural rhythm. The first may feel artificial but it really is using the essence of the cycle in a deliberate fashion.

The second thing to remember is that judgments and decisions are inextricably connected. You cannot do one without doing the other. Being aware of this will make it easier to open up to the wisdom offered by each phase of the Moon.

From the New Moon to the Balsamic, each phase describes a part of the process of judgment and how we reach emotional conclusions. The purpose here is simple: to give you a chance to know yourself and observe how you make judgments and to create the chance for you to refine or change your own process. We will examine the integral elements of each phase as well as an inherent pitfall that can keep us from using that phase to grow and evolve. In addition, a keyword is given as a focal point both for opening to the gifts of that phase and for meditation.

Take time to ponder each part of the cycle. You may gain some understanding of which facet is like a comfortable shoe and which one is like walking across nails, which one you sail through and which one sidetracks you or stalls you. Also, each of us was born under a particular phase of the Moon. If you know your birth phase, please pay special attention to it, as it may offer some added insight into how you approach judgment.

New Moon (Keyword: Engage)

This phase is all gut reaction, all instinct. The key here is to just feel. Understand that the New Moon is about first impressions. Let them sink in. No matter whether you are looking at new opportunities or responding to a crisis. Be careful though, this is the "jump-the-gun" phase. You are quick to judge and act impulsively. You go for it without taking consequences into account. If it feels right, it must be right. You are unlikely to question what you are doing and, if the consequences are not what you were anticipating, you'll find a way to rationalize your initial assessment. The reason? It really is hard to be objective because this phase is all about pure feeling.

Crescent Moon (Keyword: Entertain)

The bottom line? There's a lot going on. It's not unlike being at a crossroads with a multitude of signs pointing in countless directions. Your initial impression has given way to a whole new batch of feelings. Mix in past experiences, worries, and concerns, and it's really quite crowded in your psyche. Now is when you are likely to be deeply triggered. You may not recognize it at first because there's so much going on, but take stock. Look at your options one by one. See the opportunities. The pitfall of this phase is getting so caught up in all the alternatives that nothing takes root and you find yourself lost in a maze of unfulfilled possibilities. Forming a judgment now is premature. It likely won't stick because the purpose of the Crescent phase is to entertain ideas and seek information.

First Quarter (Keyword: Readiness)

Okay, things are starting to take shape. You sort through the input you've received and filter things out. You develop an initial assessment. Still, you are testing a decision to see if it feels right. You are defining what feels appropriate to you. What might sidetrack you here is your ego, always on high alert to defend itself. If your feelings have been hurt in any way, you might ignore everything else and act on that single reaction, short-circuiting any possible new awareness. You simply go to your default response. This is the home of the knee-jerk reaction, the part of the process where you simply don't want to get outside of your comfort zone. Let it run its course without committing you to any judgment. Then, trust yourself to step past your habitual responses and be open to what they reveal about you. Whatever you do, don't close down shop. The question now is not so much if your judgment is sound but if the timing is appropriate. It's not.

Gibbous (Keyword: Refinement)

Much of this phase revolves around asking yourself what seems to be an infinite number of questions, all of them highly analytical and often picky. Don't skip over this part! It's important to look at the direction you're headed to see if you've overlooked anything important in making an appropriate judgment. The pitfall here is obvious: Fear of making the wrong assessment can keep you hung up, going over and over the same ground until you can't see straight. It's like trying on the same dress again and again to make sure it fits. The mantra of this phase? Be prepared. Better that than taking a leap before you know where you are going to land. However, you can get so busy preparing that you never actually take the leap. Just know that you can handle whatever comes your way. And disconnect that automatic replay in your head that recycles every less-than-desirable choice you've ever made. Believe it or not, it won't help you.

Full Moon (Keyword: Clarity)

Take a deep breath. The time has come. Make a judgment. Everything is out in the open where you can see it and, more importantly, understand it. You've done everything you can to prepare for this moment and your perspective is crystal clear. No time for hesitation, just go. The path may not be smooth, but satisfaction comes from honoring yourself, your experience, and the awareness you have gained. The pitfall during the Full Moon can be procrastination, waiting for everything to be in perfect harmony and balance. The fact that you have arrived here means that things are ripe for conclusion. So, step up to the plate and make the call. It's like standing at home plate with bat-in-hand. You gotta hit the best pitch, but holding out for the *perfect* pitch may cost your at-bat.

Disseminating (Keyword: Expansion)

You may be asking, "What's next? Isn't it enough that I made a judgment that led to a decision?" Well, no, it's not. This phase is about acknowledging the change your judgment is going to make in your life. Whether that change is big or small, obvious or subtle. Every judgment leads to a shift in consciousness, especially if you have had the courage to confront long-standing issues. Now, healing and transformation can take place. Granted, you may feel uncomfortable. Welcome it. This comes from pushing yourself beyond what you know. And recognize that just under the discomfort is exhilaration. The pitfall here is refusing not to use this whole process to enhance your life, instead slipping into familiar patterns. Make no mistake, that would be like deflating a balloon. Nothing much magical or appealing about that! Not anaylyzing your judgments' affects leaves you wondering why you bothered with the judgments so much in the first place.

Last Quarter (Keyword: Reflection)

Here's where the results start to register and you get a chance to see just what you've accomplished—where your judgment has taken you, the fruits of your labor. And it's not just about whether or not you ended up where you planned. It's not unusual for life to take a left turn when you thought you were going right. Why? Life is not linear and the initial catalyst for sizing things up and making a judgment call doesn't have to directly connect to what you create. Take time for a little more judgment. Assess the whole process, where it took you, what you felt, what it showed you. Don't ignore anything. There is much gold to be mined from this. More rewards, more wisdom. Remember, every step along the journey held purpose and meaning, so don't minimize the depth of your experience. In the end, it's everything.

Balsamic (Keyword: Release)

This phase pushes you to let go of all that you have just done. Don't hang on. There is no purpose in that, only stagnation. Recognize there is no space for the new if your psyche is cluttered with the old. Allow your experience to be like compost, providing the nutrients to take on the next adventure life will no doubt present to you. Release the need to assume that the process you've just gone through can be set in stone and used forever. The Moon's cycle teaches us that everything has its day and then it is gone. What can be cherished is the continuous opportunity to experience over and over again this miraculous and profoundly practical cycle. Within it is the promise of a new self always emerging.

. . . .

The Moon really does offer us the ultimate model for implementing change, for creating growth in our lives. The impetus for it may not spring from her well, but she gives us countless openings to begin again.

So, embrace the eight phases of the Moon. Live them. Use them. They will teach you that there are no such things as mistakes, just judgments, decisions, choices that have consequences you hadn't planned on. This leads to mastery and to a new you. It may not be pleasant but it will always be instructive.

This approach can even be used looking backward in our lives. Take the model and apply it to situations you have already dealt with. See what further insights strike you. It's natural to wonder what planet you were living on when you made certain choices. In fact, it's necessary. In the wondering and the questioning, we seek to know the whys and wherefores of life, not just for entertainment but so we can improve.

Imagine what would happen if you could just celebrate the choices you've made. How much different would your life look to you? Wow, I did this and then I grew, I evolved, I became stronger. Life experiences are the raw materials each of us has at our fingertips to expand and transform, not proof that we are incapable.

A couple of other tips. It can be hard to resist the free-fall into negative judgment sometimes. But, if you find yourself wobbling on the edge, take a deep breath. Connect to your authentic feelings and then honor them by aligning yourself with what I believe is the guiding principle behind appropriate judgment: Is this judgment about making my life work better for me? Because that is the purpose of judgment—to honestly assess what is in your best interest. So, examine, examine, examine. Just do it wisely and with compassion.

Right now, we are at the tail end of a powerful opportunity to shift the use of judgment away from the critical toward a more user-friendly attitude, one that shares solutions and observations instead of pointing out faults. This shift began in September 2007 with the movement of the planet Saturn into Virgo. Saturn asks us to look at the consequences of our actions and be accountable. Virgo is the sign that always wants to know how to make things work more effectively, which is part of why its residents invented self-improvement. This combination of Saturn and Virgo has pushed us to work with issues of judgment for all of 2008 and most of 2009.

This cycle closes its limited run (it happens once every 28 to 30 years) between April and October this year. Use the energy wisely. Remember, judgment begins with the feeling world, with the Moon. Navigating this opportunity requires more than rational observation; it demands a heartfelt response based on true feeling. Go inward. Observe which phases of the Moon poke you, prod you, push you. This will provide a clue about which part of your feeling life needs healing. Use this year to clear away the clutter of negative judgments and create a new template for living and processing your life. You deserve it.

About the Author

Lesley Francis is a professional astrologer, writer, journalist, teacher, and psychic with a deep passion for and commitment to using her talents and gifts to support others in the greatest act of creativity—living their lives. Her credentials include twenty-one years at Canada's fourth largest newspaper, writing for magazines, and speaking across Canada and the United States at astrological conferences. Lesley began studying astrology thirty-five years ago, never envisioning that it would become more than an avocation. She can be contacted by e-mail at lesley_francis@hotmail.com or by going to www.andnow.ca.

Relationships in Transition

By Alice DeVille

Remember when the world anticipated the start of the new millennium ten years ago? Everyone was concerned with how our relationship to time-sensitive equipment would pan out and whether electronic systems would survive the transition. After all the handwringing over Y2K, very little occurred to disrupt the mechanics of our lives, and we continued without a hitch. As we coasted through the year 2000 and elected a new president, how were we to know that our relationship to security would be forever altered by the tragedy of September 11, 2001? Global relationships transitioned, taking on new importance; people acted differently, felt violated, and reached out to help others find hope in a world turned upside down. For many individuals, healing meant rekindling the complex relationship with living. As we head into 2010, let's take inventory of our relationships and renew our contract with balance.

What comes to mind when you hear the word *relationship*? Most individuals cite emotional or other connections between people. Others mention spouses, significant others, relatives, coworkers, and friends. Yet the American Heritage College Dictionary gives a broad definition: "The condition of fact of being related; connection or association." In 2010, a push for balance occurs when Saturn, the planet known for its restrictive energies, travels through Libra, challenging this sign of relationships to find stability while coming to grips with much-needed change. This article takes a broad look at issues affecting your personal wheel of life as you embrace an exciting new decade.

Relationship with Self

Many of us make resolutions to improve our self-image, which can be accomplished by building a better relationship with self-confidence and expression. Do you feel like others take advantage of you and you say yes when you really mean no? Make use of phrases such as, "That won't work for me," "Let me think about it and get back to you," or "I'm already booked." Perhaps being an inventive go-getter is a long-held desire to give you independence and upward mobility in the world. Create goals and a written timetable and watch the magic happen. Don't like your looks? The "new you" could be seeking enhancement from cosmetic surgery to remove the bags and jowls; laser eye surgery to ditch unflattering eyeglasses and contact lens maintenance; or a weekly Pilates session to stretch away the bulges and tone up your muscles.

Relationship with Your Consciousness

When you feel a wave of mental restlessness compounded by an urge to explore unfamiliar places, your higher mind begs for a workout. Even if you don't actually pack your suitcase and cross the ocean, give yourself a timeout to experience different viewpoints. Expansion could come from enrolling in formal classroom studies or just as easily from an online or self-study course that holds your interest. If you're opting for a change of climate, try house-sitting in another part of the country—you get free lodging while you care for the home in the owners' absence. This plan offers you a sabbatical from stale routines.

A call to learn more about other cultures may be the incentive you need to go on the dream trip, study other languages, or take a world cruise. Find the right choice among a tantalizing variety of vacation packages. Soon you'll be sampling exotic cuisine and practicing your language skills. If an interest in global humanitarian causes motivates you, find international organizations that help those suffering from disease, poverty, or political manipulation, and sign up for a working tour. Your hands-on experience is sure to be an eye-opener, giving you a keen awareness of life conditions in need of vital transition.

Relationship with Relaxation

Sometimes you go through a period in life when you feel as though you are sacrificing your essence to help others in need. Perhaps it is through visiting the sick in medical facilities, listening to a friend who needs a shoulder to cry on after a romantic breakup, running errands for elderly neighbors, doing the lion's share of carpool duty for your children and their friends, or offering comfort to those who have lost loved ones. Although you are a willing participant in giving your time and energy unconditionally, you're struggling to stay on top of your own commitments and feel like you are losing the battle with privacy. Little by little, you have agreed to the increasing demand on your time. Now the recipients of your generosity seem to take your presence for granted. Being tested in the trials of life feeds your need to escape from the burden of over-commitment.

A little tranquillity goes a long way. Start your search for it by identifying alternatives. Free up your personal time by enlisting the aid of others. Set boundaries with yourself and then share them with affected parties. The best way to transition the relationship you have with being on call 24/7 is to get away from the hectic universe. Hide out in a self-created sanctuary to recharge your batteries, meditate, read, catch up on crossword puzzles, and rekindle your visionary path. By investing in self-renewal you transcend the urgency of mundane distractions and find your inner strength. You'll understand why "Each relationship you have with another person reflects the relationship you have with yourself" (Deville, "The Moon & Relationships," *Llewellyn's 1998 Moon Sign Book*). Then go out and blaze another trail.

Relationship to Food, Health, and Nutrition

Although fat has fallen out of favor in recent decades, studies clearly show that healthy fat is important for your heart, your brain, joints, and maybe even your mood. Significant reports suggest eating more along the lines of previous generations' diets: lots of fruits and vegetables that are packed with flavor, nutrients and fiber; whole grains and dried beans; dairy; and skinless poultry and lean meat. In moderation, eat nuts and seeds; avocados; seafood; and olive, canola, and other oils rich in polyunsaturated and monounsaturated fats. Diet fads come and go (see DeVille, "Food Fads and Nutrition," *Llewellyn's 2006 Moon Sign Book*). Gram for gram, fat contains more than twice the calories of protein or carbohydrates.

No one has truly invented a way to reduce weight quickly without effort. The real secret to shedding pounds is counting calories while eating the foods your body desires. You cannot force yourself to eat what you find unappealing—that's what triggers yo-yo dieting. If you eat less than you burn, you will drop the weight; eat more than you need, and the pounds will pile on. You'll be hearing more and more about "nutrigenomics," a concept under study that examines how your genetic makeup determines what foods can best fuel you. This emerging area of research suggests that nutrition advice is likely to be more individually tailored in the future. An interesting take on going back to eating in more traditional foodways may be found in *Nourishing Traditions* by Sally Fallon. Although some critics find this book controversial, it claims to challenge the modern "politically correct" nutrition and food preparation techniques that constitute radical change from the culinary choices of our ancestors. Although there have been tremendous advances and dollars poured into research, diagnostic, and surgical specialization, degenerative disease remains on the rise. Our relationship with food is under scrutiny to determine whether products consumed contribute to emerging health problems. Let's get energized! This could be the perfect year to move our natural state of health to one of balance and vitality.

Home, Hearth, and Family Relationships

Nothing is as sacred as the space you call home and the people you call family. Ideally, home is where you feel anchored and secure. Who knows you more intimately than your family of origin? Your surroundings should be welcoming and relaxing. When they're not, it's time to assess the cause and take action to change the dynamic. Your age and status in life often drive how well you handle emotional exchanges. Are you getting along with family members? Do you have toxic sibling relationships? How well do you respect others' boundaries? The more self-sufficiency you demonstrate, the more you prepare for emancipation. As tension sets in, your mind works overtime with the recurring thought that maybe it's time to leave the nest and set up housekeeping on your own.

Perhaps you need a bigger nudge. Mom and Dad think it's time you flew the coop and hope you get the hint: they raise your share of room and board, Mom announces she's planning a redecorating project—starting with your room. There's nothing like new restrictions to get you thinking about your desire for independence! If you have outgrown your digs and have the money to branch out, why not move? Will you miss the "womb service" and the home-cooked meals? It won't take long before you are self-sufficient—don't let fear hold you back. You're bound to find your purpose in life by gravitating toward a new beginning and appreciation of self.

Many adults face entirely different challenges in relationships as parents age and need medical care and assistance in finance and housing. A parent may come to live with you, or you may help them downsize and move into more manageable living quarters. Single adults frequently move back home to care for ailing parents. When there are other siblings, it's always a good idea to make sure everyone is aware of parental needs and preferences as well as the cost of elder care. Smooth communication helps you avoid squabbles down the road when major decisions must be made to allocate assets or ask for help. And if you are the one approaching retirement age, it is never too early to plan for how you wish to live out your golden years.

Significant Partnerships

Have you ever wondered why your restless spirit drags its feet when it is time to assess relationships? Sometimes, the more intimate the relationship, the more difficult it becomes to initiate communication when something feels unsettled. My article in *Llewellyn's 1998 Moon Sign Book*, "The Moon & Relationships: Taking Inventory of Emotional Needs," discusses compatibility with those you have drawn into your circle and encourages you to find the right fit when your connections fall short of the mark. The article covers close contacts in love, friendship, and business. With the planet Saturn moving through the relationship sign of

Libra in the next few years, many of these associations will be put to the test. Some will survive; others will dissolve. In this segment, let's dwell on intimate partnerships, such as marriage partners, lovers, or significant others.

Evaluating relationships begins with acknowledging your feelings about love and trust. When you feel strain and distance in another's presence when you used to feel joy and anticipation, you may begin to question commitment. Initially, you may judge yourself harshly for entertaining doubts. You or your partner may be in denial that a problem exists, at least for a while. Gradually, new patterns in the relationship replace those you cherished and either complacency or suspicion evolves.

Discovery of Affairs

When your significant other changes routines so drastically that you can't count on your partner showing up for dinner or attending planned events, it's time to ask questions. Don't hold your breath hoping to get truthful answers right away; mainly, you will fuel your suspicions. Those who don't want to get caught philandering have mastered juggling techniques that work for them for years on end. The ones who feel the most guilt start leaving trails: unaccounted for receipts, suspicious phone calls or phone numbers, scheduling gaps, and stories that just don't jibe. Before long, all the raw emotions surface—anger, hurt, or revenge. The couple may decide on counseling or reject it altogether. If a state-recognized relationship is beyond repair, a legal intervention is likely.

Ending Relationships

If this is a divorcing couple or a pair with shared assets, ending the relationship is complicated. Both are in need of healing, yet practical issues—separation agreements involving property, custody arrangements if there are children, and attorney representation—receive the most attention. How nice would it be to look at the failed relationship as part of the growth cycle rather than a stigma? It works best to call a truce, look at the terms of the partnership and living arrangements, and figure out a way to cope until one or both parties can make a move. The process of rebuilding self-esteem and changing the status quo would then be more tolerable and easier on all parties.

Neighborhood and Community Relationships

What can you give back to your community? If you always feel "too busy" to donate time for causes, take another look at your calendar and consider building these relationships. How well do you know your neighbors and the local leaders and organizers? A helping hand goes a long way to make the environment hum. Try volunteering for initiatives that support the homeless, the elderly, the sick, children, or the homeowners' association. If you write, maybe you can develop the community newsletter and focus on events, movers and shakers, heroes, school news, green environment projects, or scout activities.

Having problems getting along with neighbors? These are vital relationships. Where you live affects the quality of daily living. Life could be teaching you to absorb what your senses perceive with an air of caution. Maybe your communication style needs work. What we learn from close encounters brings out the need for diplomacy and adaptability. If you are fighting over parking spaces, boundary issues, the misplaced fence posts, unsightly border shrubs, or the barking dogs that wake you far too early on weekend mornings, it's time to sharpen your negotiation skills. What can you do? Tackle the problem by ringing the offender's bell with a smile, a plate of baked goods, and a sincere attitude toward resolving the issue. To break the ice, compliment your neighbor on the new paint job or other home improvement you've noticed. Then be direct in broaching the sore spot and ask what "we" can do to solve it. You may even be able to muster up a sense of humor over the inconveniences endured and wind up being good friends with the individuals you thought of as the "nasty next-door neighbors."

Are you a frequent host? Be sure to let neighbors know when you are going to have a large get-together that could absorb much of the available parking space. Even better, invite these neighbors over for a barbecue or a holiday party and spread a little cheer. You'll have the first word on being a good host and the last word on community harmony.

Recreation and Social Activity

Come up for air and live a little! This statement takes aim at self-proclaimed workaholics who put tasks at the top of the to-do list and vacation at the bottom. If that description fits you, how long has leisure time been taking a back seat to pressing deadlines? Those of you with a strongly developed restrictive zone may be uncomfortable taking a day off or putting a long-desired dream

2009 © Andres Rodriguez. Image from BigStockPhoto.com

trip on your agenda. Perhaps you derive satisfaction and enjoyment from more purposeful enterprises and avoid opportunities to let your hair down and relax. You may not even have a clue how to behave in a less formal environment. I once had a relationship with someone who wore a dress shirt and slacks to the beach—that not-so-subtle message took the fun out of enjoying the salty air and sandy shores.

A colleague recently told me that his wife booked a one-week anniversary trip to Jamaica and wanted the trip to be about them. Actually, she said it was about her—him spending the week adoring and pampering her without distractions. She spelled out conditions: no laptop, no Blackberry, and no cell phone turned on to take those "urgent" business calls while the couple enjoyed the resort amenities. As I write this section, the duo is away, and I can't wait to hear how it all turns out.

Your relationship to social activity ties directly to your ability to enjoy yourself through all forms of pleasures, amusements, and games. Include some form of exercise in your recreation and you automatically improve your health. By participating wholeheartedly in adventurous undertakings and stretching your perspective, you'll easily earn a reputation for balancing work and leisure time, a much-desired quality for those on the executive track.

Relationship to Your Career and Work

In many industries, downsizing due to diminishing consumer demand leads to layoffs, demotions, or a shift in the actual components of the job. You may have wanted to stay with your employer for life, but the company solvency dictates otherwise. It's always wise to keep your resume up-to-date so you can start applying for new positions. Perhaps you find yourself in a situation where you no longer relate to the philosophy of the workplace, and relationships are strained. These conditions validate your suspicions and set the stage for your next step: a visit to the headhunter.

To enjoy a quality of life that matches your goals, pursue a new job or angle for a raise. Do whatever it takes to earn more to keep up with inflation. If a part-time job helps you whittle away debt, go for it. Take classes or enroll in certification programs that give your resume a boost. Cultivate mentors to advise you on strategic career moves and pursue your dream job with confidence.

Relationship to Organizations

If you have ever been in a situation where you are a dues-paying member of an organization but haven't attended a meeting in years, this section is for you! Group relationships often run their course. Formal associations draw you in when you relate to a common cause. You may even hold an office for a while and become one of the organization's movers and shakers, doing your best and giving your all to promote the mission. Then the dynamic shifts and you find yourself looking for excuses to avoid the next meeting. A leadership change or a shift in goals could be responsible for your lack of interest. Perhaps previous employment responsibilities drove your attraction to a professional group but your new job is not a good fit. It's okay to let yourself off the hook and resign.

Are you drawn to future-oriented objectives and want to connect with kindred spirits? That's easy—expand your networks. Take a risk by accepting invitations to introductory meetings. If you like what you see and hear, join. You'll form new associations and set new goals. Before long you'll expand your "family" of friends.

Relationship with Resources

The relationship you have with your money can get an overhaul thanks to the rising cost of living and shrinking dollar values and equity investments. In all likelihood, you have rearranged your spending priorities and are looking for ways to pummel the pain at the gas pump by ditching the gas-guzzler and getting cozy in a smaller, energy-conscious vehicle, maybe a hybrid. Never able to stick to a budget before? Examine your money habits or you'll wind up scratching your head trying to figure out where the money went since you began using credit cards in lieu of cash. Want to test the new approach to purchasing power? Maximize cash flow by dining out at a nice restaurant every other week instead of weekly. Carry a nutritious lunch to work in lieu of chowing down on the $10-a-day deli lunch. Drink home brewed instead of gourmet coffee for a month and see how much you can save—experts say you'll net more than $1,500 if you do this for a year. Promote environmental friendliness by using filtered tap water instead of buying water in plastic bottles.

Caving in when you shop for groceries? Resist the urge to stockpile on every bargain you see—the same item will go on sale in a few weeks when you really need it. Make a list and stick with it when you shop. Avoid aisles with tempting treats that add unplanned dollars to your grocery bill and unwanted pounds to your waistline. Coupon clipping is on the rise; consumers report savings of more than $20 per shopping trip.

If it sounds like you're going to develop a poverty complex, scrap that thought. This advice is about making conscious rather than robotic choices.

Relationship to Investments and Debt

In addition to your relationship with money in general, you also maintain relationships with the money you've saved and the money you owe: your assets and debts. A major concern in many households is the rising cost of debt and the diminishing value of assets. Escalating credit card debt is a prevailing financial problem calling for immediate attention and a debt-reduction plan. Those who thought making the mortgage payment with a credit card was a good way to rack up bonus miles are having second thoughts. While they used to pay the card balance in full each month, these consumers find they no longer have disposable cash and use credit cards to pay for groceries, gasoline, and staples. A big complaint is that no money goes into savings. This scenario calls for drastic transformation of your monetary picture. What do you do?

Draw up a budget, following these steps: 1) Purchase a basic financial software program and list your assets, accounts, income sources, and liabilities. 2) Figure out your income but leave out any unexpected bonuses or awards. 3) Track your expenses, including monthly bills and everyday purchases. Note months when larger one-time payments, such as taxes or insurance, occur. 4) Do the math to see your bottom line and keep a worksheet handy. 5) Once you have your ideal monthly budget, stick to it! The bottom line: a return to financial health!

About the Author

Alice DeVille is an internationally known astrologer, writer, and consultant. In her busy Northern Virginia practice, she specializes in relationships, government affairs, career and change management, Feng Shui, real estate, and business management. She works under contract with the federal government as an executive coach, management consultant, and writer. Alice is also a licensed Realtor with a number of real estate credentials, including financial management expertise. She has developed and presented more than 160 workshops and seminars related to astrological, Feng Shui, metaphysical, motivational, real estate, and business themes. Alice's writing appears on the StarIQ.com Web site and other nationally known venues, including Learning Escapes. Her work has been translated in numerous languages, appearing on Web sites, newsletters, and publications in the United States and abroad. Alice's Llewellyn material on relationships has been cited by Sarah Ban Breathnach in Something More *(Grand Central Publishing, 2000), Oprah's Web site, and in material used by The Federal Executive Institute and The Franklin Planner (a line of day planners). Alice focuses on helping clients discover the spiritual, psychological, and practical tools that support their life purpose. Contact her at DeVilleAA@aol.com.*

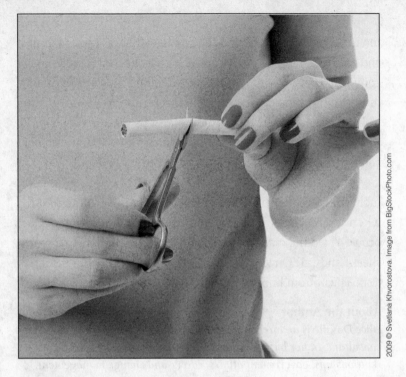

Improving Your Habits with Lunar Help

By Maggie Anderson

Almanac devotees are familiar with the formidable list of agricultural and farm husbandry chores that can be done successfully under the right Moon phase. These include planting potatoes, making cheese curds, and cutting hay. I keep my almanac handy just in case that five-pound bag of potatoes under my sink begins sprouting and demands to be planted. The almanac's listings of Moon's phase and the zodiac sign for each day are useful for all sorts of practical matters. For instance, if I substitute the word *grass* for *hay*, I know when it's time to mow my lawn!

After Bibles, almanacs were the second publication to roll off the newly invented European printing presses around CE 1439. Society was primarily rural then and had a long tradition of coordinating agricultural activities with the phases of the Moon. Our agricultural ancestors had health and grooming concerns similar to our own: that's why we've inherited their knowledge about the best phases of the Moon to have haircuts, visit the dentist, and begin a new business.

Although our forefathers and mothers had the same physical, moral, and spiritual strengths and weaknesses as we do, they didn't emphasize personal development as much as building character and community. Personal goals were not an issue for them and the phrase "self-development" wasn't used often—if ever—in the earliest almanacs.

However, it is a huge waste of moonbeam power to ignore the Moon and her phases when pursuing a primary focus of modern humans: self-improvement. Luna can provide the cosmic support needed to give up bad habits or incorporate good ones into our daily lives. Almanacs are not just for growing potatoes and going to the barber anymore: we can use their information to locate the correct lunar energies and gather cosmic support for personal enhancement projects.

Resolutions to be better humans are not just for New Year's Eve anymore. Some goals of modern men and women have developed into full-blown self-help industries, complete with personal coaches, television programs, famous authors, and publishing houses. They fall into two main categories.

Break Bad Habits at the Full Moon

The first category of goals focuses on problematic behaviors that should be decreased or abandoned completely for our own well-being or the well-being of our families and the planet. Curiously, it is easier to notice these in other people than in our own lives.

Some habits have become perennial favorites: smoking, overeating, addictions to chocolate and high fructose corn syrup—as well as caffeine, alcohol, various drugs—and using plastic bags.

Self-improvement gurus also target the overuse of television and the Internet ("Help us!!! We can't stop blogging!!!"), and cell phones that are glued to our ears ("Help us!! We can't stop talking!!!"). Former President George W. Bush told Americans that we are addicted to oil ("Help us! We can't quit driving!!!).

There are also intangible addictions—the ones that make elderly relatives nervous and drive roommates crazy. We always promise them we'll quit. These might include impersonating politicians, beat boxing, and watching *King of the Hill* reruns.

We can and should tap Luna's assistance when giving up negative habits. The formula is simple and easy to remember, too: to eliminate a negative behavior, stop doing it on the Full Moon, when the Moon is decreasing. Throw away your cigarettes, cigars, lights, matches, and even those 1960s collectible ashtrays when the Moon is full. Do not ritually bury them in your backyard because it will be too convenient to dig them up again at 3:00 am tomorrow morning, by the light of the Full Moon. Stop addictive behaviors during the Full Moon and they will eventually disappear from your memory bank, not easily, but much easier than you might expect.

Once those negative habits have been purged, they can be replaced with positive ones. Forming new positive habits keeps us from being bored with all the free time we have because we no longer have to stand in snowbanks to smoke cigarettes in January or play computer solitaire for hours on summer days in June.

Adopt Healthy Habits at the New Moon

The second variety of self-improvement goals are those we believe make us better individuals. These aspirations usually include daily exercise and reinforcing our bodies with a healthier diet.

Setting and reaching career goals, obtaining financial security, athletic excellence, having a great marriage, and being a superior parent are also popular self-development goals. It's a long list: we can use the help of the Moon to become this super-achiever. (Though a personal trainer wouldn't hurt, either.)

Following that must-do list, add-ons fall into the "I must develop my God-given talents" category and include learning to speak fluent Chinese and climb Mount Everest. It's best to focus on only one of these at a time—Moon energies are powerful but there's only so much that can be accomplished in a few weeks.

Finally, there is a category of goals labeled "I want to become more spiritual and, overall, a better person." The path of this goal includes meditation, journaling, fasting, and other spiritual disciplines promoted by faith communities.

Begin practicing the new behaviors when the Moon is New. Often, positive goals are attempted in the same time period as we are decreasing the bad behaviors. Realistically, it's better to concentrate on giving up one negative behavior first. That's enough to deal with. Start adding positive replacements about six weeks after the "quit date" for a negative habit, and do it at the New Moon.

It's an Interesting Idea, But Why Should I Do It?

The basis for coordinating activities with the phases of the Moon comes from this simple but profound belief: life events go more smoothly when coordinated with the cycles of nature. We already follow the cycles of the Sun. Humans sleep better at night than during the day, as anyone who works second or third shift will attest. With the exception of those hardy souls that mark the New Year by jumping into icy lakes, most of us know we'll have a better, more relaxing swim when snowmen are not watching.

Going a step further is giving attention to the cycles of the Moon, which is not that much of a stretch from planning activities around the sunshine and seasons. Moon phase electional astrology does require a bit of research and planning, but the technique is simple: look in this almanac to find the phase of the Moon, and coordinate your habit-breaking and -forming goals to the signs and phases. The best way to find out if the Moon phase approach "works" in your life is to try it!

Refining the System

To add a cosmic layer of insurance to your next self-improvement project, consider the keywords associated with each zodiac sign.

If the habit you want to increase falls under the influence of one of the signs below in the "Increases" column, note the associated zodiac sign. Consult the almanac and find the next New Moon in that sign. Mark the date as the right one to begin new behaviors.

Is there a behavior you'd like to discontinue? Check the "Decreases" column below for a match. Look through the almanac for the Full Moon in the related zodiac sign. Mark your calendar for that date as the best "quit date."

May the Moon shine on all of your efforts!

	New Moon Increases	**Full Moon Decreases**
Aries	Independence	Bad habits in general
Taurus	Spirituality, financial success	Rigid thoughts and behaviors
Gemini	Learning, adaptability	Smoking, lack of concentration
Cancer	Family ties, emotional maturity	Overeating, alcohol use
Leo	Self-mastery, creativity, romance	Drug use, gambling
Virgo	Efficiency, good nutrition	Worry, negative self-talk
Libra	Cooperation, social ties	Conflict, social obstructions
Scorpio	Mutual trust, exchange	Score-keeping
Sagittarius	Active lifestyle, good study habits	Unfocused activities
Capricorn	Career goals, maturity, wisdom	Procrastination, work problems
Aquarius	Humanitarian goals, teamwork	Emotional detachment
Pisces	Charity, psychic abilities	Sleep disorders, escapist behavior

About the Author
See page 298 for Maggie Anderson's biography.